SCIENCE COMICS

THE BRAIN

The Ultimate Thinking Machine

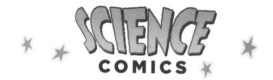

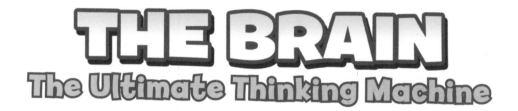

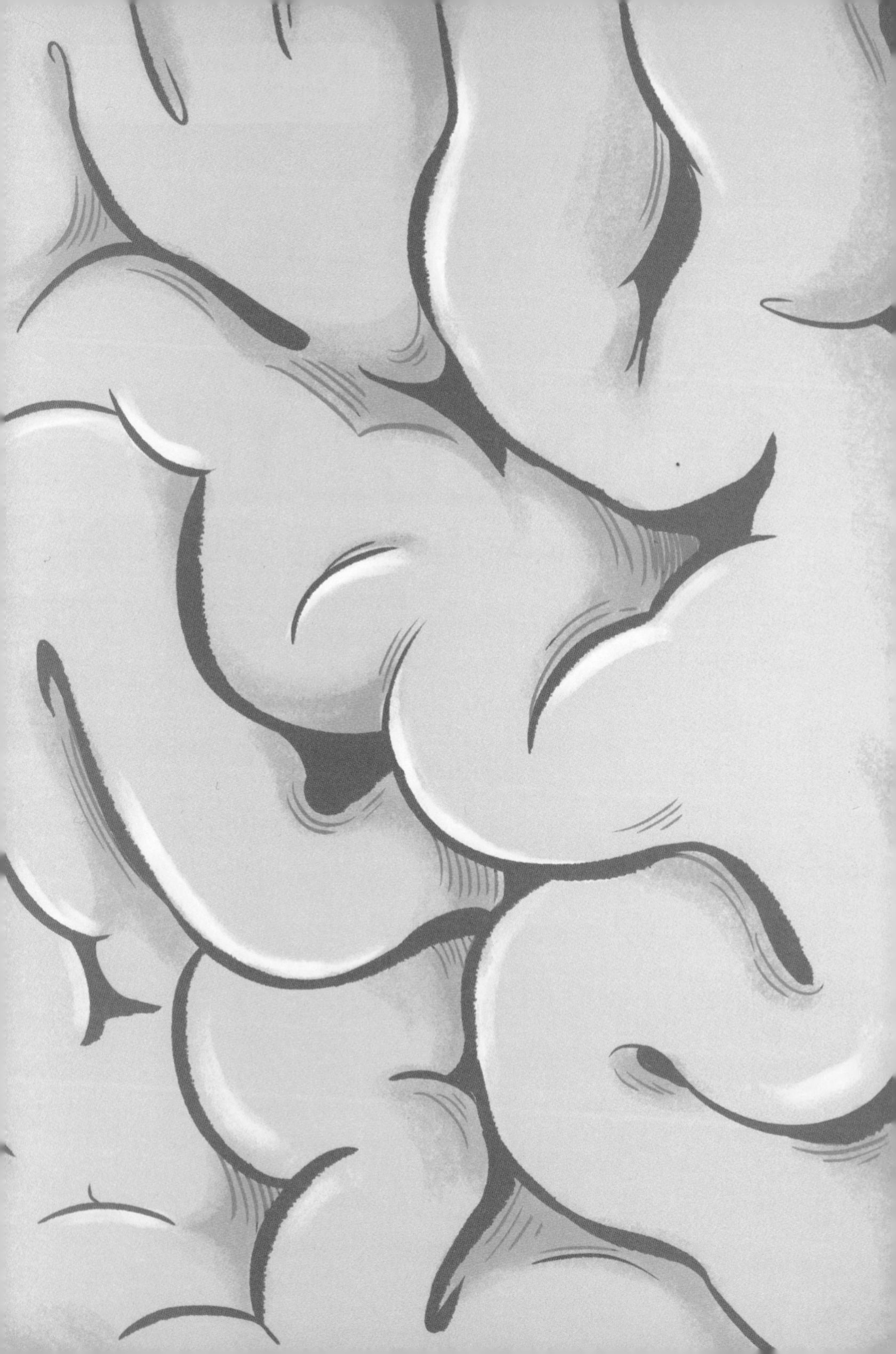

THE BRAIN
The Ultimate Thinking Machine

written by
Tory Woollcott

illustrated by
Alex Graudins

SCHOLASTIC INC.

For my mom and dad, who gave me this great brain.
And my granny, who showed me how to use it!
—Tory

For Mama and Papa Graudins for their endless support.
—Alex

Edited by Dave Roman.
Book design by John Green.
Brain consultant: Jordan Poppenk.

Drawn and colored with Adobe Photoshop CS5 on a 22HD Wacom Cintiq.
Lettered with Comicrazy font from Comicraft.

Text copyright © 2018 by Victoria Woollcott.
Art copyright © 2018 by Alex Graudins.
All rights reserved. Published by Scholastic Inc., 557 Broadway, New York, NY 10012,
by arrangement with Roaring Brook Press, an imprint of Macmillan Publishing Group, LLC.
Printed in the U.S.A.

ISBN-13: 978-1-338-76299-0
ISBN-10: 1-338-76299-0

4 5 6 7 8 9 10   40   29 28 27 26 25 24 23

Scholastic Inc., 557 Broadway, New York, NY 10012

My love of science first began with an appreciation of the stars in the sky. And I fell in love all over again when I discovered the stars in our minds.

When I was twelve years old, my grandpa gave me a book called *Cosmos*. I loved the stories the author, Carl Sagan, shared about our universe—the billions of stars in our sky, and the millions of galaxies beyond our own. Thinking about the vastness of space made me think about our lives here on Earth, all the things we've accomplished, and what it means to be human. How did we create spaceships to take us to the moon? How did we build cities? How do we write books, and draw comics, and create art and science? How did we come up with language? How do we *think?*

Your brain is like a galaxy inside your head, with a hundred billion stars called *neurons*. Each brain is like an incredibly powerful computer. Every one of those hundred billion neurons can make up to a thousand connections with other neurons. That means there are over a hundred trillion connections in the brain—more than a thousand times as many stars as scientists think there are in our Milky Way galaxy! Like the wires sending information inside your computer, these connections send information all over your body to keep you walking and talking. These special, precious cells—and their helpers, called *glia*—are what make you, YOU! When you were born, you already had almost all of the neurons you'll ever have. They were just waiting to grow and adapt as you got older, learned new things, and became your own person.

We still know very little about our universe, and we know even less about the brain! Every day, neuroscientists, psychologists, and cognitive scientists chip away at tiny questions to try to better understand our minds. Scientists are using all kinds of incredible tools and technologies to try to figure it out— like changing the DNA of cells to make them glow in rainbow colors under a microscope, or shining lasers on neurons that have special light-activated proteins to control the brain! Still, we're nowhere *near* being able to answer all the big questions. Questions like: What is a memory? Where does personality come from? Are there any other animals with minds like ours? How can we heal the brain after an injury, or prevent the diseases that destroy it?

These unknowns were part of what made me fall in love with neuroscience. As a neuroscientist, I get to ask my own questions about the brain: How do all of our brain cells work together to help the brain grow and develop? How do the proteins made by different kinds of cells affect the ways that neurons connect and communicate with one another? How do they go wrong in diseases? Like all scientists, I hope that, someday, I'll know the answers to some of those questions—and I'll share those answers with you.

In the next pages, you'll get to meet two sisters, a mysterious mad scientist, and his zombie butler who will all work together and illustrate the things we've learned in neuroscience so far. While you read, spend some time thinking about *thinking*. Isn't it crazy that you're using your brain to read a book about the brain, learn new things about the brain, and store that information in your brain so you can retrieve it later and share it with someone else—so they can repeat the whole process with their *own* brain? Given how mind-boggling it is to even try to imagine *that*, maybe it's not a surprise that we still know so little about the brain!

For now, enjoy learning about what we *do* know—and by the time you've finished reading, scientists may have learned something new!

—Alison Caldwell, aka Alie Astrocyte,
Writer and Host of Neuro Transmissions,
a YouTube channel all about the brain,
PhD Candidate of Neuroscience, UC San Diego, CA, USA

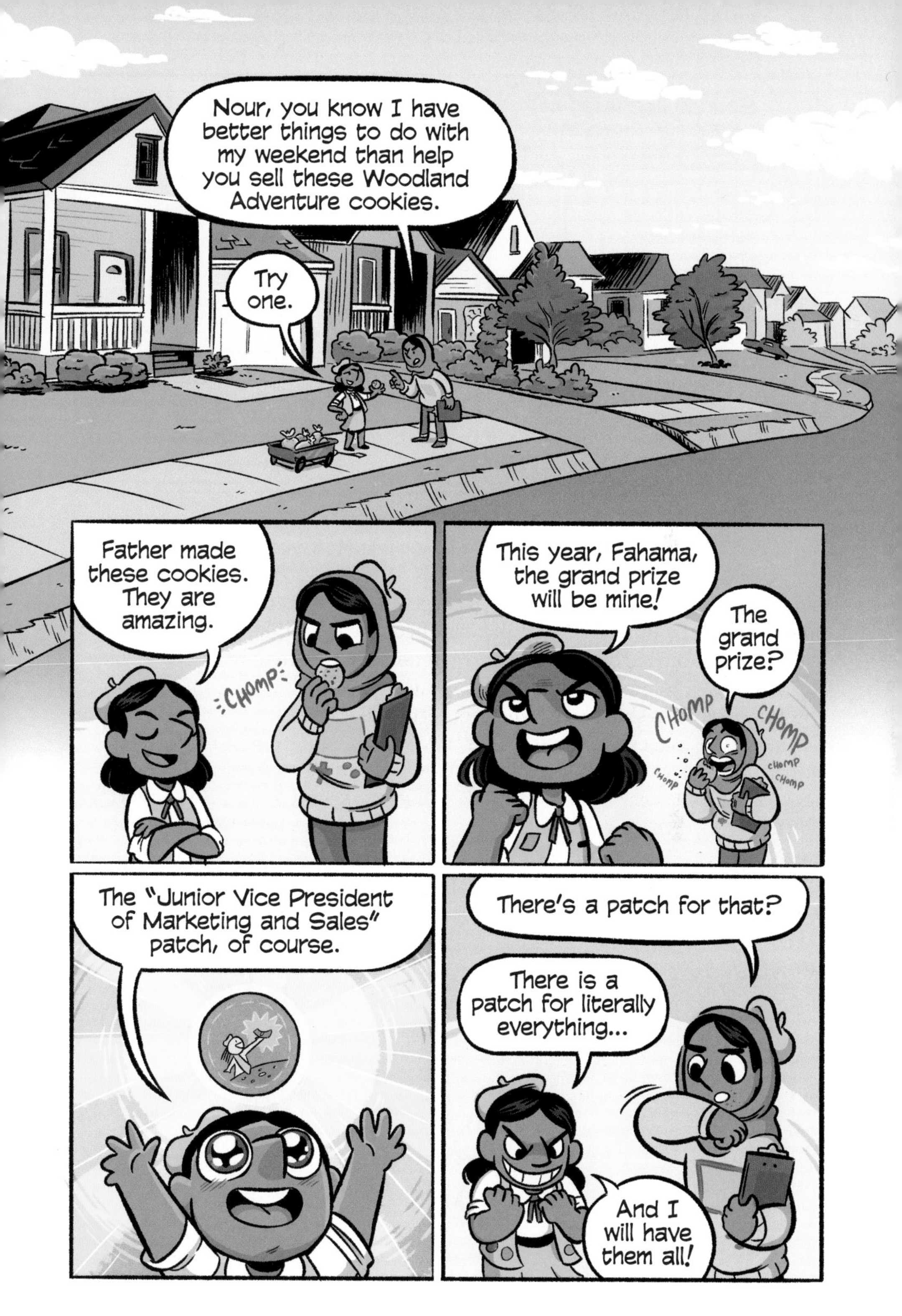
Nour, you know I have better things to do with my weekend than help you sell these Woodland Adventure cookies.
Try one.
Father made these cookies. They are amazing.
CHOMP
This year, Fahama, the grand prize will be mine!
The grand prize?
CHOMP
CHOMP
CHOMP
CHOMP
CHOMP
The "Junior Vice President of Marketing and Sales" patch, of course.
There's a patch for that?
There is a patch for literally everything...
And I will have them all!

Sounds like a lot of work for a patch.

The patch comes with a dumb video game system or something.

Video game? What system?

Help me sell the most cookies, and maybe you'll find out.

Okay, okay. So what's the plan?

We will crush our enemies and see them shattered before us.

You mean Troop 12?

I *destroyed* Troop 12 last year, and they are no longer a threat to me now! They are still trying to piece together their scattered sales strategy!

MWAHAHAHA!

I will destroy them all.

We are going to split up! I will take this block. You take that one. We'll meet back here in an hour.

Nour, you know I'm supposed to stay with you.

But we'll sell more cookies if we split up!

Don't you want that video game system?

Well...

Do you know what *I* want? I want my sister to be happy!

...Happy to *SELL THESE COOKIES!*

Okay, see you in an hour, then!

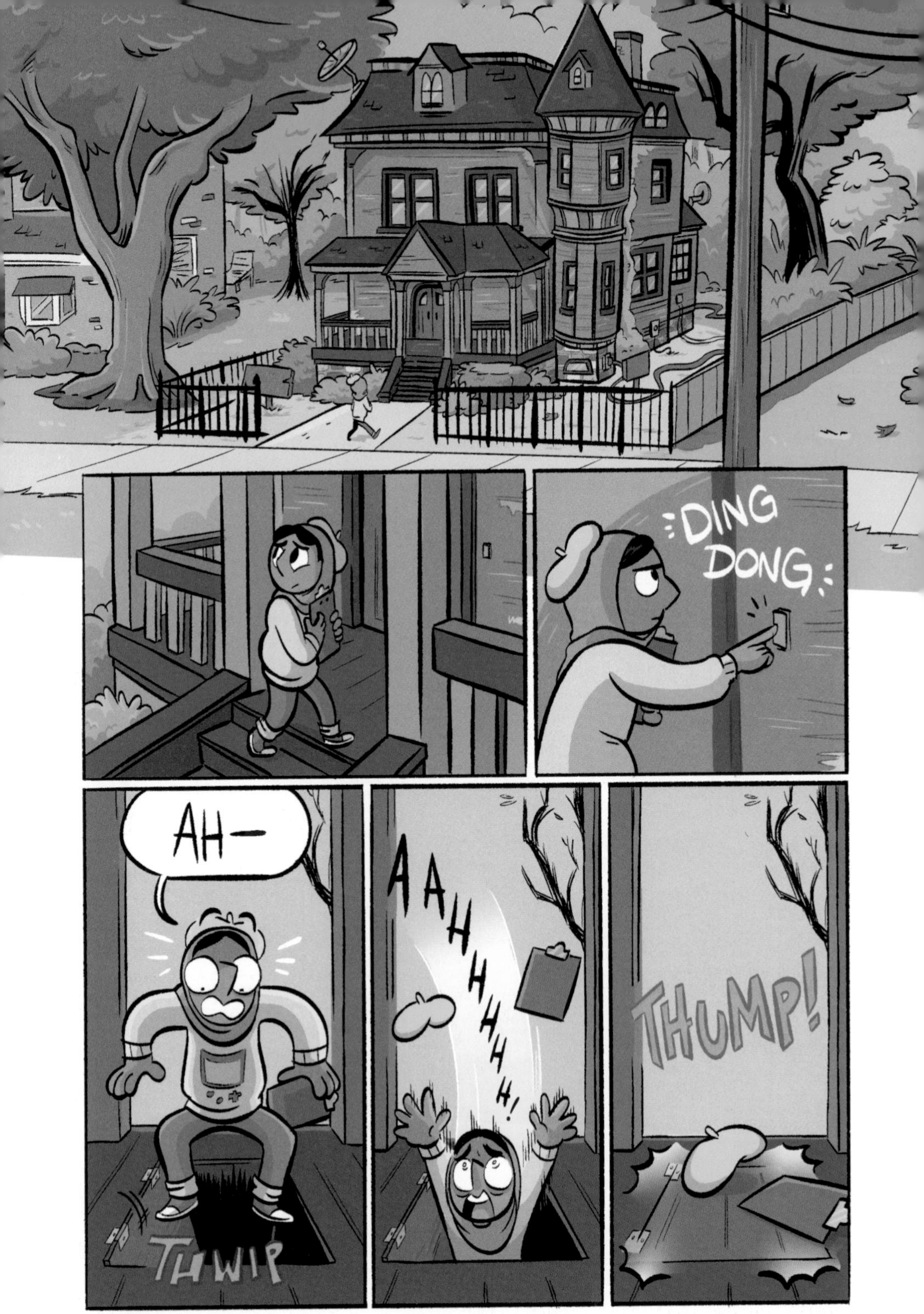
DING DONG
AH—
AAHHHHHH!
THWIP
THUMP!

UGH.
?

H-hello?

Is anyone there?

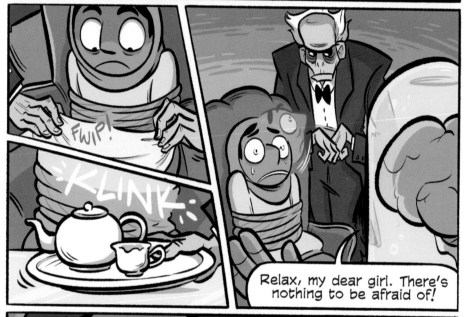

FWIP!

·KLINK·

Relax, my dear girl. There's nothing to be afraid of!

AAHHHH!

Hello, child! I am Dr. Cerebrum, and I will be removing your brain for science today!

Wh-what?

No, *no*, *no*. It'll be fine. You'll like it! Brain science has come a long way, you know.

I'm...very confused right now.

Allow me to explain! Better yet, let me take you on a *JOURNEY*...

...through the *MIND!*

Wait, what?!

No, my dear, not *what*...*WHEN!*

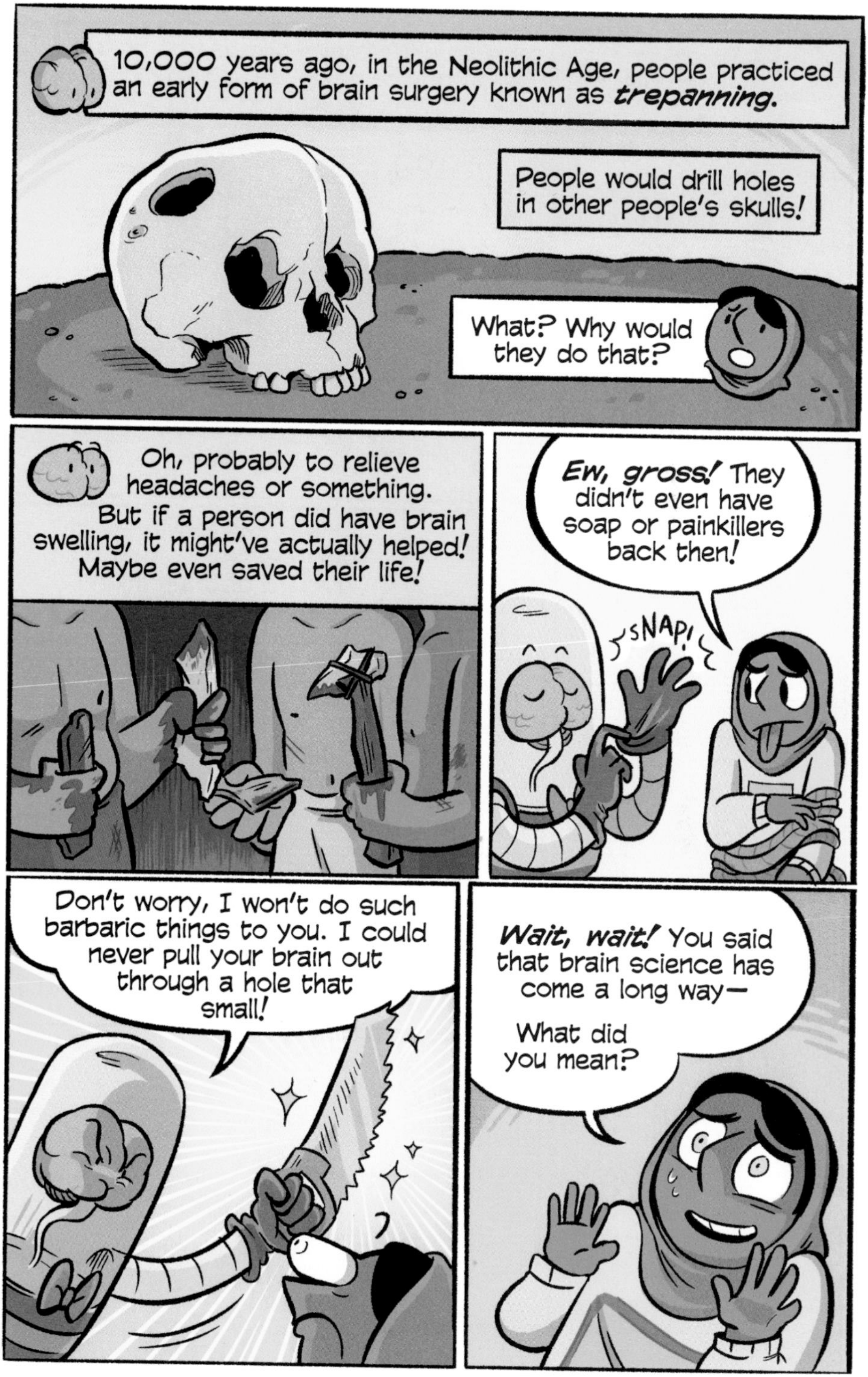
10,000 years ago, in the Neolithic Age, people practiced an early form of brain surgery known as *trepanning.*
People would drill holes in other people's skulls!
What? Why would they do that?
Oh, probably to relieve headaches or something. But if a person did have brain swelling, it might've actually helped! Maybe even saved their life!
Ew, gross! They didn't even have soap or painkillers back then!
SNAP!
Don't worry, I won't do such barbaric things to you. I could never pull your brain out through a hole that small!
Wait, wait! You said that brain science has come a long way—
What did you mean?

In science, you have to understand the basics first! Each discovery builds on the next, leading to more insight. You can't run until you've learned how to walk!

It's the same in biology! For most of history, people thought that the heart was the seat of knowledge, not the brain.

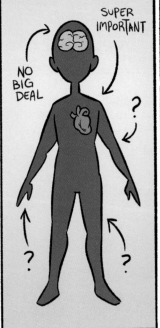

SUPER IMPORTANT

NO BIG DEAL

?

?

?

This is the **Edwin Smith Papyrus**, the first written record of brain injury!

Archaeologists believe it was written by Imhotep, famous healer and architect of the first pyramid.

Ancient Egyptians knew that if someone had a head injury, that person would behave strangely, but they didn't know why.

It's just a gaping head wound, nothing to worry about.

The Egyptians thought so little of the brain, they didn't even keep it when they mummified their dead! They kept all the important stuff in special containers called **canopic jars.**

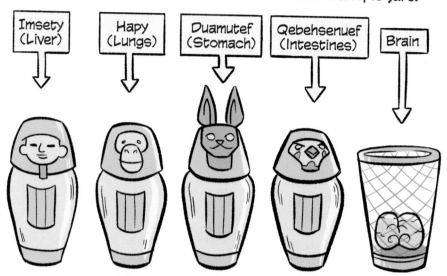

Imsety (Liver)

Hapy (Lungs)

Duamutef (Stomach)

Qebehsenuef (Intestines)

Brain

Contrary to popular belief, the Egyptians didn't actually pull out the brain through the nose but through the back of the head, via the foramen magnum!

SHOOMF!

And don't even get me started on what they used to think the brain was made out of!

It's really gross.

I'm good, thanks.

NUDGE!

Well, let me tell you anyway...

The Greeks actually thought the brain was made out of phlegm!

The Chinese believed the brain was made out of bone marrow!

BONE MARROW

The Egyptians thought the brain was made out of mucus!

MUCUS

So, wait...everyone thought the brain was made out of snot?

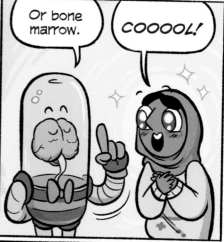

Or bone marrow.

COOOOL!

Speaking of cool...
The Greek philosopher Aristotle believed that the heart was where thought occurred and the brain was just there to cool your blood, like an air conditioner for your body!

If the head is injured, then the brain won't be able to cool the blood, and the person will behave strangely!

Meanwhile...

Hmph. Of course my lazy sister isn't here yet.

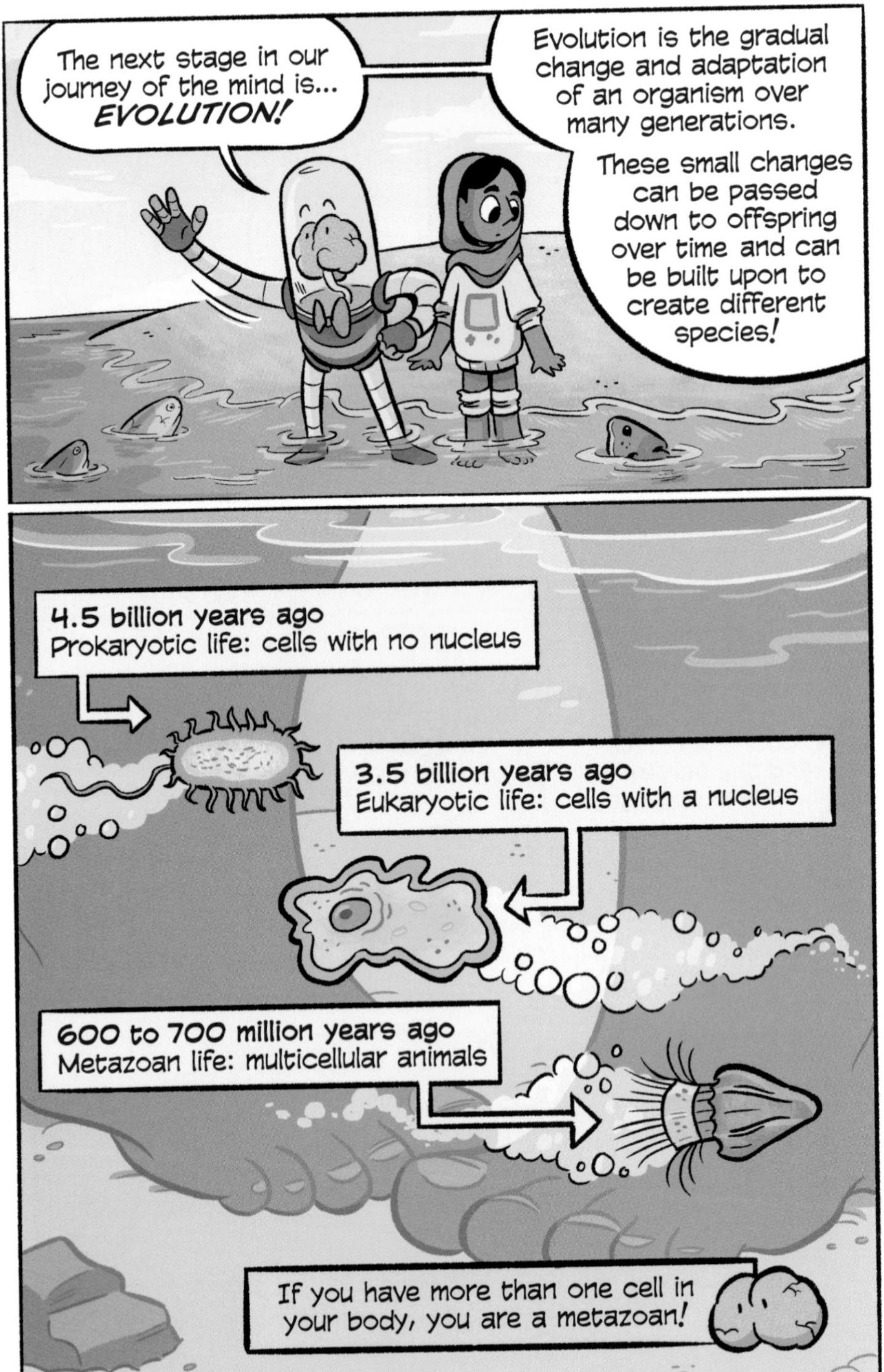

The next stage in our journey of the mind is... *EVOLUTION!*
Evolution is the gradual change and adaptation of an organism over many generations.
These small changes can be passed down to offspring over time and can be built upon to create different species!
4.5 billion years ago
Prokaryotic life: cells with no nucleus
3.5 billion years ago
Eukaryotic life: cells with a nucleus
600 to 700 million years ago
Metazoan life: multicellular animals
If you have more than one cell in your body, you are a metazoan!

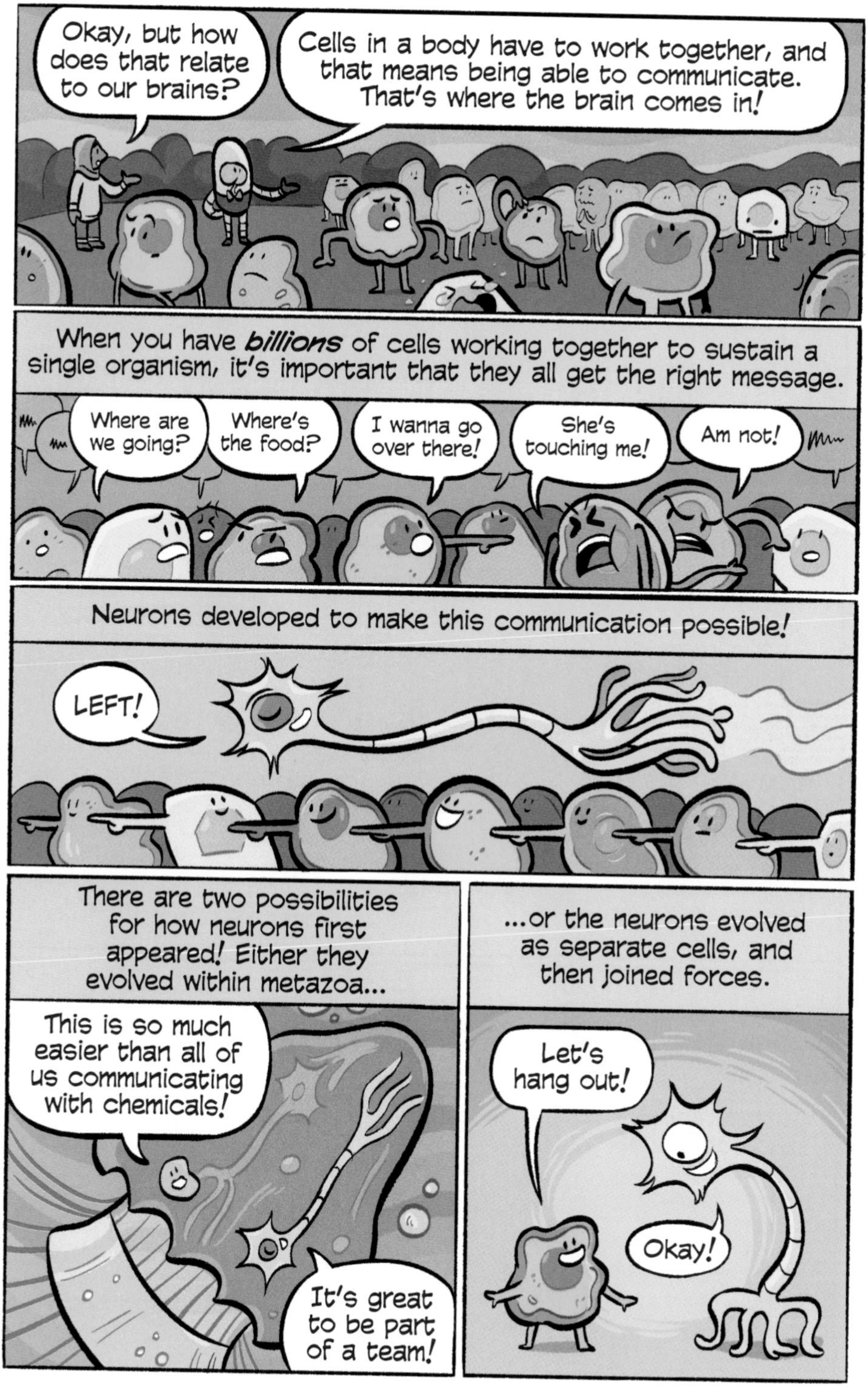

Okay, but how does that relate to our brains?
Cells in a body have to work together, and that means being able to communicate. That's where the brain comes in!
When you have *billions* of cells working together to sustain a single organism, it's important that they all get the right message.
Where are we going?
Where's the food?
I wanna go over there!
She's touching me!
Am not!
Neurons developed to make this communication possible!
LEFT!
There are two possibilities for how neurons first appeared! Either they evolved within metazoa...
This is so much easier than all of us communicating with chemicals!
It's great to be part of a team!
...or the neurons evolved as separate cells, and then joined forces.
Let's hang out!
Okay!

But just because you have neurons doesn't mean you need a brain! When neurons started working together, **nervous systems** began to develop!

Suddenly life went ballistic! New forms of life started to appear, because cells could now work as a team!

These new types of nervous systems can be separated into two groups:

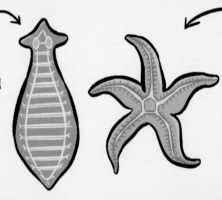

Bilateral Symmetry appears in organisms that are divided into two symmetrical halves, left and right. Humans, birds, and worms are all examples of bilateral symmetry.

Radial Symmetry appears in organisms that don't have a distinct left and right side. Sea anemones, jellyfish, and starfish are all examples of radial symmetry.

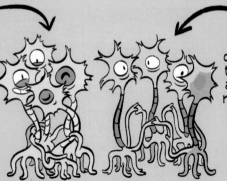

Cephalization is when a group of neurons gather in one place at one end of the body and form a bump called a ganglion. This is a huge step for evolution!

Radial symmetry led to the development of the **nerve net**, a collection of neurons spread throughout an organism's body that functions like a brain.

14

Fish

During the Cambrian explosion, about 530 million years ago, fish developed the first true brain and spine. A fish's central nervous system mostly controls movement and deals with the senses!

Amphibians

About 370 million years ago, in the Devonian period, lobe-finned fish evolved into early amphibians. Their brains were very simple, and had more in common with their fish ancestors than with other land animals!

KEY

- ▨ **Olfactory bulb:** used to detect smell
- ▨ **Forebrain:** controls behavior
- ▢ **Midbrain:** used for motor control
- ▨ **Hindbrain:** controls automatic body functions

Reptiles

Reptiles evolved from amphibians around 310–320 million years ago, during the Carboniferous period.

Birds

During the Triassic period, over 230 million years ago, dinosaurs evolved! Unfortunately, we don't have any dinosaur brains lying about, but we do have their descendants—birds!

Up to that point, brains were small and incredibly specialized. **Mammals** added the neocortex on top of the existing brain, which had already been evolving for so long, making it more adaptable.

With the arrival of primates, the neocortex grew bigger, causing the frontal lobes of the primate brain to grow larger as well!

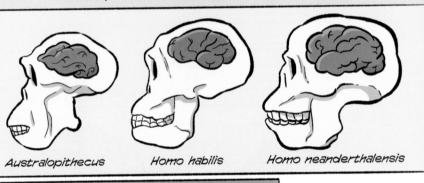

Australopithecus   *Homo habilis*   *Homo neanderthalensis*

Around 800,000 years ago, there was another sudden jump in brain size. This was when primates related to humans arrived—**Neanderthals!**

Wait, aren't Neanderthals like cavemen?

You say that like it's a bad thing.

Technically, Neanderthals had a larger brain than modern humans.

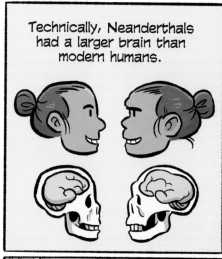

But to be honest, I doubt you'd be able to tell the difference between a Neanderthal and a short, muscular modern person.

Oooh.

Thank you for your time, ladies.

TP TP TP TP TP

SLAM!!!

Fahama! Where are you? Fahamaaa!

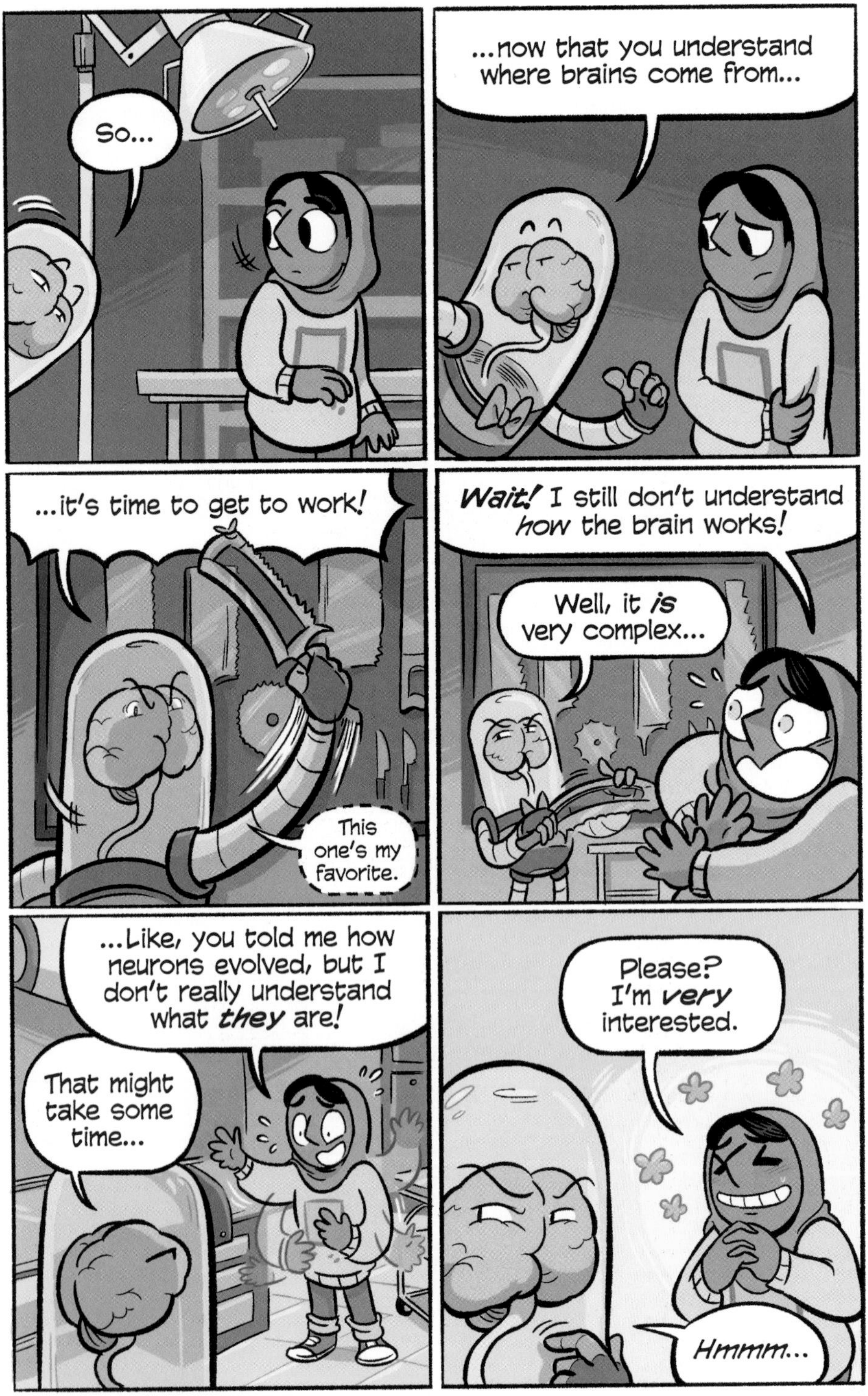
So...
...now that you understand where brains come from...
...it's time to get to work!
This one's my favorite.
Wait! I still don't understand *how* the brain works!
Well, it *is* very complex...
...Like, you told me how neurons evolved, but I don't really understand what *they* are!
That might take some time...
Please? I'm *very* interested.
Hmmm...

Meanwhile...

Mother.

It is my theory that a rival Woodland Adventure troop has abducted my sister, in the foolish belief that it will curtail my cookie sales.

That's nice, dear.

Mother, you underestimate the threat Troop 7 pose. They are a worthy adversary. I admit, I hadn't considered the possibility of abduction... If this was their plan, they are truly *diabolical*.

Uh-huh.

Fear not, Mother. I shall investigate.

Junior P.I.

Well, now that we've covered the brain's evolution, we can get on to the good stuff!

Braaaiinss...

TOSS!

Yes, yes, my dear chap. All in good time.

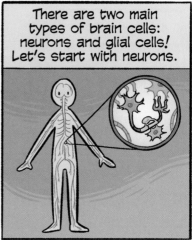

There are two main types of brain cells: neurons and glial cells! Let's start with neurons.

Neurons are cells that can send messages through chemical and electrical signals. They are capable of moving muscles, interpreting stimuli, learning—even creating thought itself!

...−−−...

In order to understand neurons...

...we need to first talk about regular cells.

Cells are the basic unit of life—your body is made up of billions of cells!

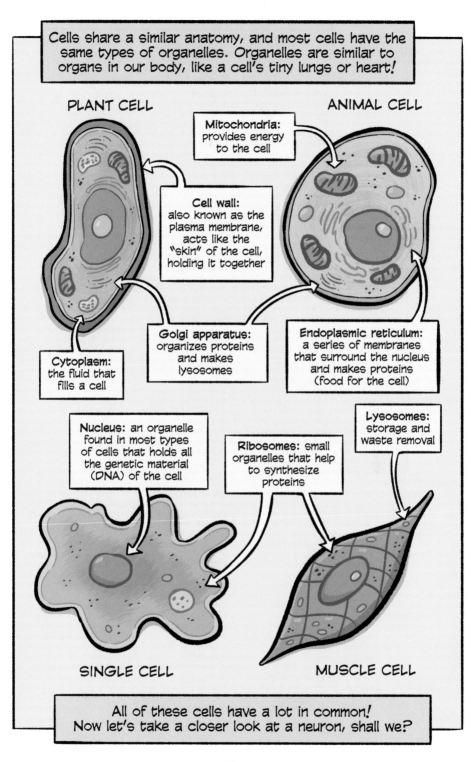

Cells share a similar anatomy, and most cells have the same types of organelles. Organelles are similar to organs in our body, like a cell's tiny lungs or heart!

PLANT CELL

ANIMAL CELL

Mitochondria: provides energy to the cell

Cell wall: also known as the plasma membrane, acts like the "skin" of the cell, holding it together

Golgi apparatus: organizes proteins and makes lysosomes

Endoplasmic reticulum: a series of membranes that surround the nucleus and makes proteins (food for the cell)

Cytoplasm: the fluid that fills a cell

Lysosomes: storage and waste removal

Nucleus: an organelle found in most types of cells that holds all the genetic material (DNA) of the cell

Ribosomes: small organelles that help to synthesize proteins

SINGLE CELL

MUSCLE CELL

All of these cells have a lot in common! Now let's take a closer look at a neuron, shall we?

Neurons have a unique shape.

'Sup?

STOMACH CELL

LIVER CELL

SKIN CELL

NEURON

Neurons also make electricity!

You're so cool!

SKIN CELL

NEURON

I'm glad you agree.

The first thing that makes **neurons** unusual is that they don't reproduce.

I reproduce often in order to grow the body or replace damaged cells!

REGULAR CELL

NEURON

I hardly reproduce at all! Human bodies are born with most of the neurons they'll ever need, and we're well protected, so we rarely need to be replaced!

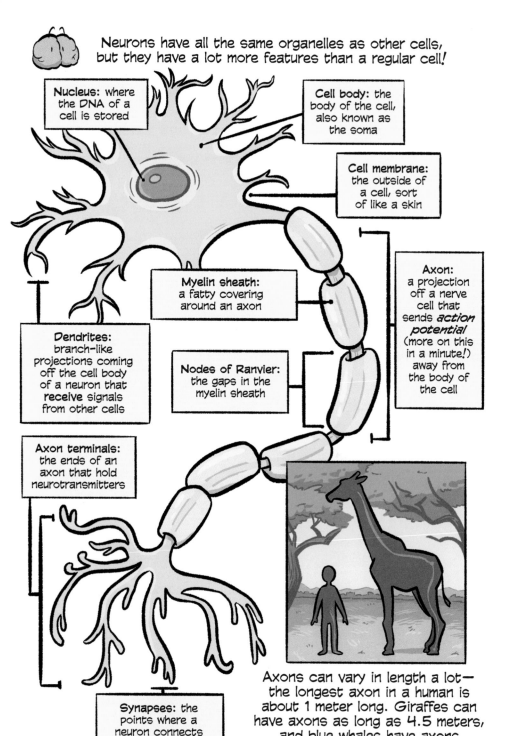

Neurons have all the same organelles as other cells, but they have a lot more features than a regular cell!

Nucleus: where the DNA of a cell is stored

Cell body: the body of the cell, also known as the soma

Cell membrane: the outside of a cell, sort of like a skin

Axon: a projection off a nerve cell that sends *action potential* (more on this in a minute!) away from the body of the cell

Myelin sheath: a fatty covering around an axon

Dendrites: branch-like projections coming off the cell body of a neuron that **receive** signals from other cells

Nodes of Ranvier: the gaps in the myelin sheath

Axon terminals: the ends of an axon that hold neurotransmitters

Synapses: the points where a neuron connects to another cell

Axons can vary in length a lot—the longest axon in a human is about 1 meter long. Giraffes can have axons as long as 4.5 meters, and blue whales have axons up to 20 meters in length!

Members of the same cell type will often look the same. A skin cell from your face will look very similar to a skin cell from your hand or your butt!

Neurons are different! Neurons can come in many different shapes. They all have the same basic structures, but can look surprisingly different from one another.

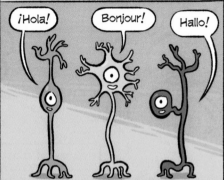

A **bipolar cell** is a type of neuron that has two long branches projecting out of the cell body. They are used for sending out specialized signals for the senses.

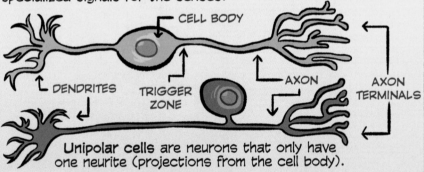

CELL BODY

DENDRITES

TRIGGER ZONE

AXON

AXON TERMINALS

Unipolar cells are neurons that only have one neurite (projections from the cell body).

Multipolar cells are neurons that have a single axon. They also have a lot of dendrites so that they can receive action potentials from the axons of many other neurons.

MYELIN SHEATH

Okay, so what is this "action potential" you keep talking about?

Action potential is what we call the electrical communication between cells!

A neuron's job is to communicate with other neurons and cells throughout the rest of the body. A neuron generates electrical signals along its cell membrane and inside its cell body. These signals are called **action potential**.

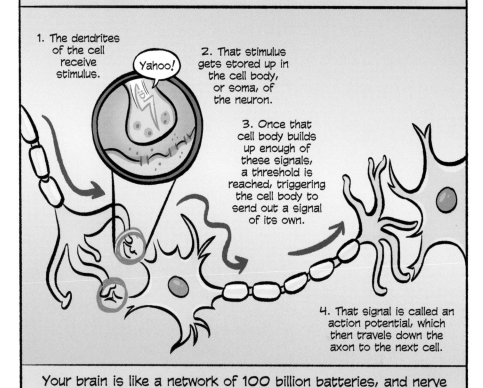

1. The dendrites of the cell receive stimulus.

Yahoo!

2. That stimulus gets stored up in the cell body, or soma, of the neuron.

3. Once that cell body builds up enough of these signals, a threshold is reached, triggering the cell body to send out a signal of its own.

4. That signal is called an action potential, which then travels down the axon to the next cell.

Your brain is like a network of 100 billion batteries, and nerve signals can move at speeds of up to 540 kph (335 mph)!

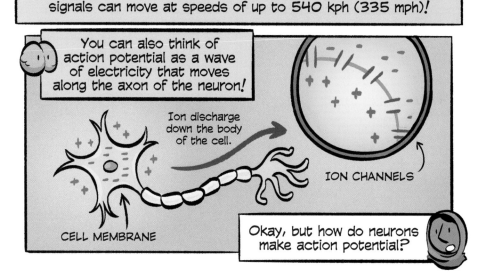

You can also think of action potential as a wave of electricity that moves along the axon of the neuron!

Ion discharge down the body of the cell.

ION CHANNELS

CELL MEMBRANE

Okay, but how do neurons make action potential?

Ah! When you look at a neuron, the outside of the cell membrane is covered in positive ions, while the inside is filled with negative ions.

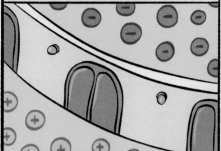

When the neuron is stimulated, little channels open that let the positive and negative ions swap.

The swapping of positive and negative ions is what triggers action potential!

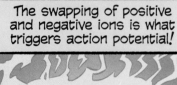

This conversion of chemical energy into electrical potential—the action potential—is what passes down the axon of the neuron to the axon terminal.

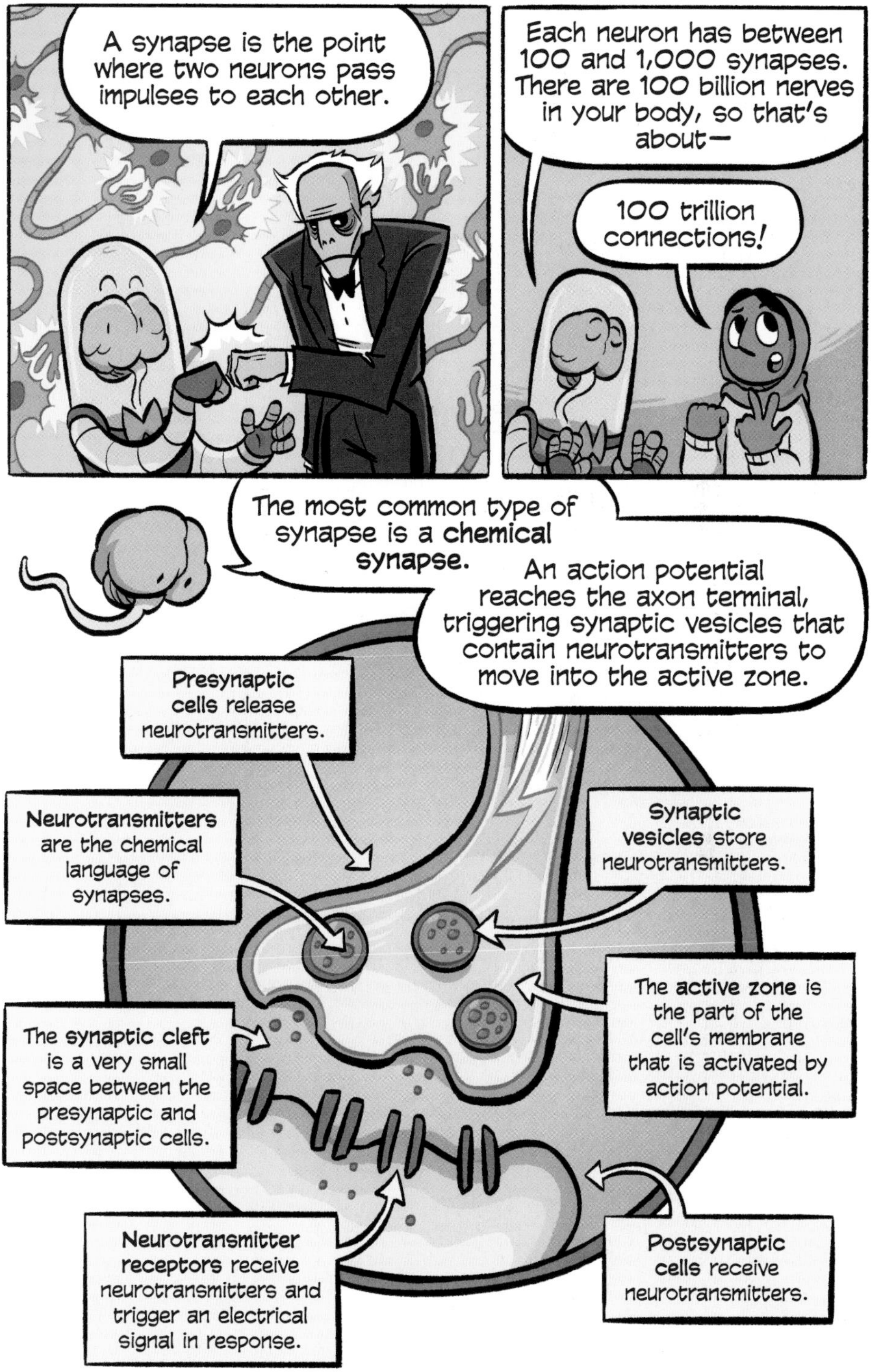
A synapse is the point where two neurons pass impulses to each other.
Each neuron has between 100 and 1,000 synapses. There are 100 billion nerves in your body, so that's about—
100 trillion connections!
The most common type of synapse is a **chemical synapse.**
An action potential reaches the axon terminal, triggering synaptic vesicles that contain neurotransmitters to move into the active zone.
Presynaptic cells release neurotransmitters.
Neurotransmitters are the chemical language of synapses.
Synaptic vesicles store neurotransmitters.
The **active zone** is the part of the cell's membrane that is activated by action potential.
The **synaptic cleft** is a very small space between the presynaptic and postsynaptic cells.
Neurotransmitter receptors receive neurotransmitters and trigger an electrical signal in response.
Postsynaptic cells receive neurotransmitters.

Neurons talk to each other via chemicals called neurotransmitters, which are released by axon terminals and are passed to the receptor of a neighboring cell. Each receptor can only react to one type of neurotransmitter.

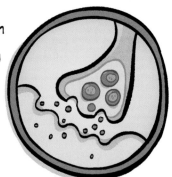

Because each axon is connected to so many other neurons by its axon terminals, it's important to be picky about which neuron gets which message, otherwise things get confusing very quickly.

In effect, neurotransmitters filter out all cells the message isn't for!

Each type of neurotransmitter is like a key that can only open one receptor type— one "door," if you will.

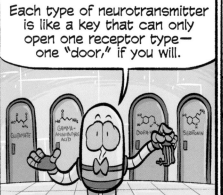

When a "key" is put into its matching "door," the neuron will then unlock smaller doors on the cell membrane. If enough doors are opened, the cell will fire an action potential.

DOPAMINE

Neurotransmitters are grouped into two main types: excitatory and inhibitory.

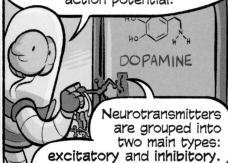

Excitatory neurotransmitters are neurotransmitters that make a neuron more likely to fire action potential.

Inhibitory neurotransmitters are neurotransmitters that make a neuron less likely to fire an action potential.

There are more than 40 different types of neurotransmitters.

Glutamate is the most common excitatory neurotransmitter.

Seratonin

Gamma-aminobutyric acid is the most common inhibitory neurotransmitter.

Dopamine

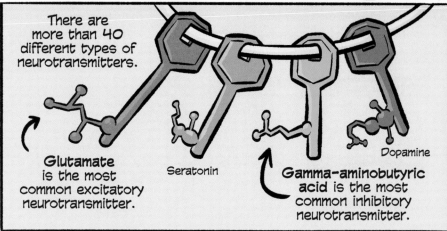

Non-neuron cells found in the brain are called *glial cells*.

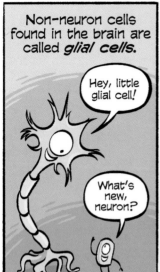

Hey, little glial cell!

What's new, neuron?

Unlike neurons, glial cells don't make electricity. But there are ten times more glial cells in the brain than there are neurons.

Hi! Hi! Hi! Hi! Hi! Hi! Hi! Hi! Hi! Hi! Hi!

There are four main types of glial cells:

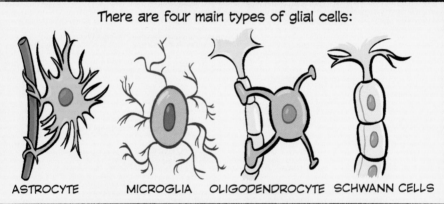

ASTROCYTE　　MICROGLIA　　OLIGODENDROCYTE　　SCHWANN CELLS

Each of these cells has a distinct appearance, and each has a totally different role in the nervous system!

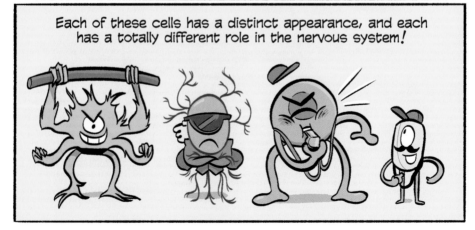

Astrocytes are the biggest and hardest-working cells in the nervous system!

I guess you could say astrocytes are the real *stars* of the show!

Get it? Because they're shaped like stars?

Anyway, astrocytes provide nutrients and structural support to neurons, which is super important!

This is key to how the neuron, its synapses, and your brain grows and develops!

ASTROCYTE CELL BODY

CAPILLARY

NEURON

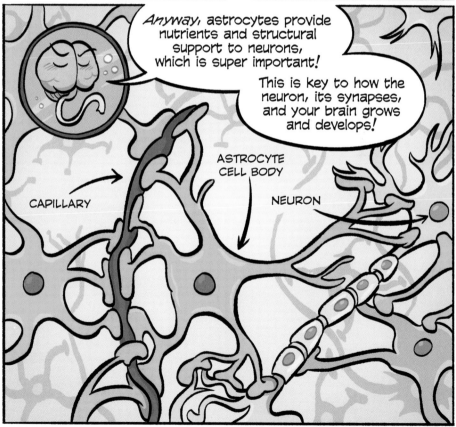

Microglia are specialized immune cells only found in the brain! About 10–15% of all the cells in your brain are microglia.

Microglia protect your brain from intruders, and they also keep the place clean.

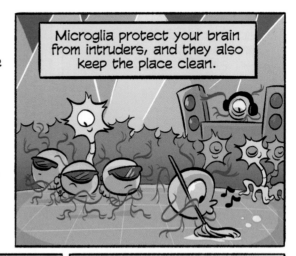

These cells are active! They move around the brain a lot, and they change shape in order to do this.

In order to be able to move between tightly packed nerves in the brain, microglia squish themselves very thin and pull themselves around.

The microglia become blobs when they need to absorb and remove "garbage" from the brain.

STOP! INTRUDER!

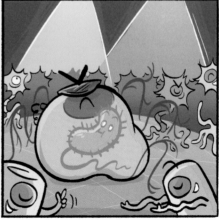

Oligodendrocytes and Schwann cells are pretty similar! They both wrap themselves around the axons in a process called myelination, which creates the **myelin sheath.**

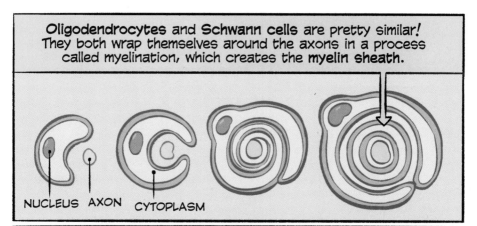

NUCLEUS AXON CYTOPLASM

The oligodendrocytes and Schwann cells support and protect the long, delicate tendrils of the axons projecting off the nerves.

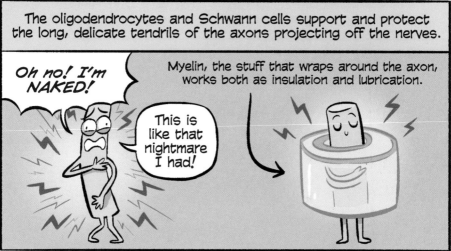

Oh no! I'm NAKED!

This is like that nightmare I had!

Myelin, the stuff that wraps around the axon, works both as insulation and lubrication.

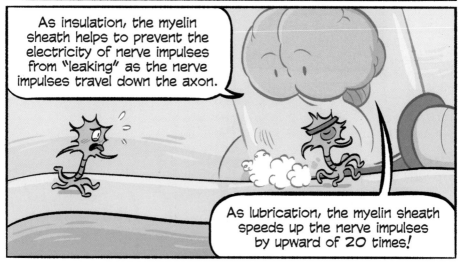

As insulation, the myelin sheath helps to prevent the electricity of nerve impulses from "leaking" as the nerve impulses travel down the axon.

As lubrication, the myelin sheath speeds up the nerve impulses by upward of 20 times!

So then, what *are* the differences between oligodendrocytes and Schwann cells?

The main difference is that oligodendrocytes are found in the central nervous system and can support and protect up to 50 axons at a time!

FEEL the BURN

HUFF

HUFF

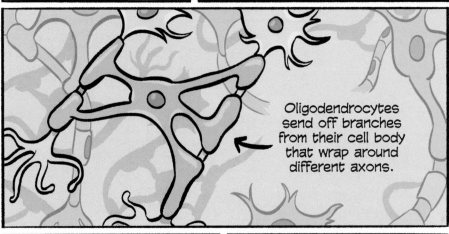

Oligodendrocytes send off branches from their cell body that wrap around different axons.

On the other hand, Schwann cells are found in the peripheral nervous system and only wrap around one axon.

In fact, there can be hundreds, even thousands, of Schwann cells on a single axon!

Elsewhere in the neighborhood...

Hi! Would you like to purchase some homemade cookies to send kids like me to camp?

Sure, give me two bags.

245

Nice, we might win this thing!

Stop right there, Troop 7 scum!

What have you done with my sister?!

Trying to distract us, Nour? *No dice!* I'm going to win the "Junior Vice President of Marketing and Sales" patch...

...and there's nothing you can do to stop me!

As the Woodland Adventure rulebook clearly states: "No Woodland Adventurer shall knowingly and with forethought distribute non-lovingly-and-homemade cookies for sale, on pain of..."

"...*FORFEITURE OF UP TO FIVE PATCHES!*"

NOOO! Anything but that! We'll tell you *EVERYTHING!!!*

Okay, so the brain is the control center of the body, but there *must* be more to it than that!

Oh yes! SO MUCH MORE!

The brain is the most complicated thing in the known universe!

There are more connections in your brain than there are stars in the Milky Way!

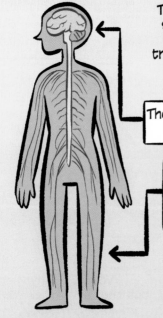

The brain is a part of the **nervous system**, and the nervous system is the network of neurons that transmit signals through your body! The nervous system is made up of two parts:

The **central nervous system** is the brain and spinal cord.

The **peripheral nervous system** is made up of all the neurons that send information to the brain and help control your body!

The nervous system is responsible for three functions: **sensory functions**, **motor functions**, and **integrative functions**.

Sensory functions gather sensory information about the world around us, like touch and smell, but also internal information like "Am I hungry?" "What part of my body hurts?"

SNIFF

GRUMBLE

Motor functions are used any time you move! (Which is a lot, I suppose, if you're cursed with a human body.)

TAP TAP TAP

Integrative functions interpret the information from your sensory functions and tell the motor functions what to do. These functions make the decisions as to what needs to be done. In other words, "thinking!"

So now that you know the nervous system, let's cut yours out!

N-not YET! I don't even know what the nervous system is made up of!

The peripheral nervous system's main job is to connect the central nervous system to the rest of the body.

I'm like a translator.

Making sure everyone gets the message!

There are two parts to the peripheral nervous system: the somatic nervous system and the autonomic nervous system.

The **somatic nervous system** controls the voluntary movements of your skeletal muscles—any muscles that are attached to a bone! It also signals pain, like when you stub your toe!

The **autonomic nervous system** controls all the nonvoluntary or self-regulating movements of your internal organs and glands, like your heartbeat and digestion! It also tells you stuff like if your stomach is upset!

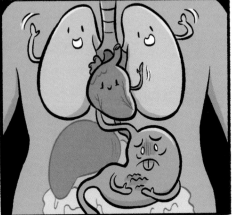

The **sympathetic nervous system** is part of the autonomic nervous system and is designed to kick in when you're in danger.

Fight or flight: this system kicks in if you're scared or stressed out—like if you're being chased by a lion!

The **parasympathetic nervous system** is the chilling system—it brings the body back to a point of homeostasis, especially after something scared you.

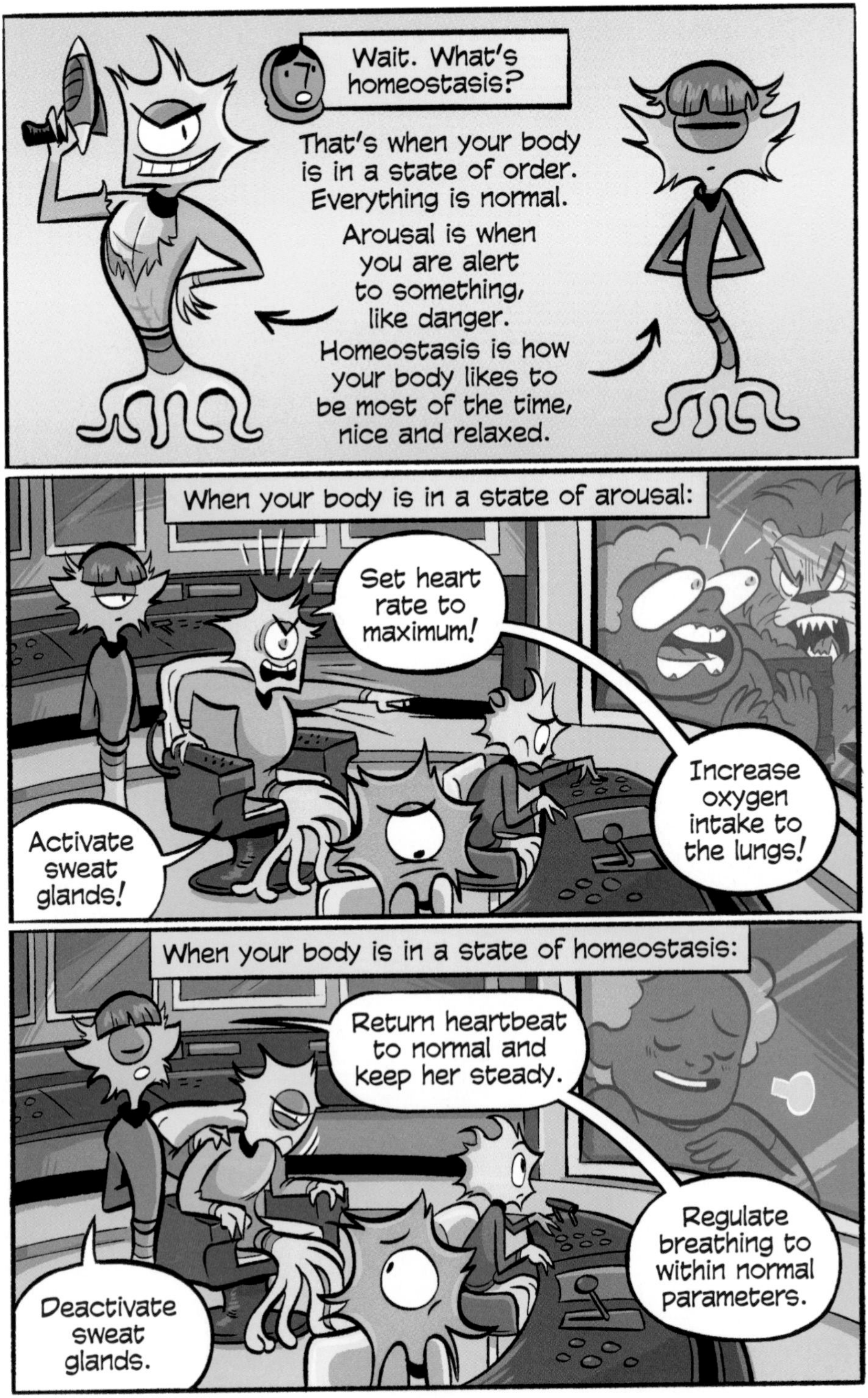

Wait. What's homeostasis?
That's when your body is in a state of order. Everything is normal.
Arousal is when you are alert to something, like danger.
Homeostasis is how your body likes to be most of the time, nice and relaxed.
When your body is in a state of arousal:
Set heart rate to maximum!
Increase oxygen intake to the lungs!
Activate sweat glands!
When your body is in a state of homeostasis:
Return heartbeat to normal and keep her steady.
Regulate breathing to within normal parameters.
Deactivate sweat glands.

The *central nervous system* is made up of both the brain and spinal cord, but let's start with the spinal cord! It's what connects the brain to the peripheral nervous system.

There are four main spinal nerve groups: the cervical nerves, thoracic nerves, lumbar nerves, and sacral nerves.

Each one of these gives you sensation, feeling, and movement to a particular part of your body!

Cervical nerves: These are the nerves in your neck! They're in charge of your arms, neck, and upper body.

Thoracic nerves: These are the nerves in your upper back and lower body.

Lumbar and Sacral nerves: These nerves are located in your legs and the organs below your belly button.

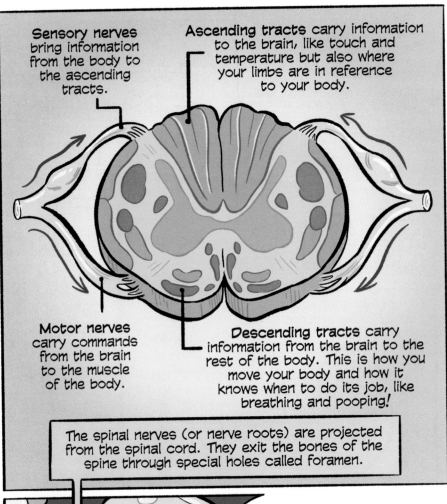

Sensory nerves bring information from the body to the ascending tracts.

Ascending tracts carry information to the brain, like touch and temperature but also where your limbs are in reference to your body.

Motor nerves carry commands from the brain to the muscle of the body.

Descending tracts carry information from the brain to the rest of the body. This is how you move your body and how it knows when to do its job, like breathing and pooping!

The spinal nerves (or nerve roots) are projected from the spinal cord. They exit the bones of the spine through special holes called foramen.

The spinal cord has three main functions:

1) It's how sensory information passes from the peripheral nervous system to the brain.

2) It's how commands get from the brain to the rest of the body.

3) It's a reflex center.

Mother, after my extensive interrogation of Troop 7, I am no closer to finding my sister. But the investigation continues...

Here, sweetheart, have a cookie!

Who made this? It smells of burning.

I helped! Your dad had to take a call.

It's gingerbread! It tastes better than it smells!

This cookie tastes of burning *and* lies.

BLECH!

No one appreciates true genius!

The brain is super important, obviously! And because most neurons cannot easily be replaced if damaged, the body takes extra precautions to keep it safe!

The skull provides structure and protection for your brain. The skull is 6.5–7.1 millimeters thick, and that's not even including the muscles, skin, and hair on top.

The bones on the skull are the first line of defense! There are eight separate bones that grow together as you age.

The skull is made up of eight bones:

2 PARIETAL BONES

1 SPHENOID BONE

1 FRONTAL BONE

1 ETHMOID BONE

1 OCCIPITAL BONE

2 TEMPORAL BONES

The skull can protect your brain from day-to-day falls or minor impacts, but your skull can't win a fight with a car or concrete! Wear a helmet and protect that beautiful, precious brain of yours!

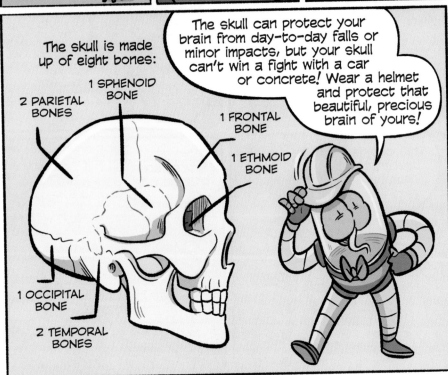

After the skull, the next line of defense is the blood brain barrier!

The blood brain barrier is pretty much what it sounds like—it's a border that stops bad things from entering your brain.

But there are things from the blood that your brain does need, like oxygen and nutrients, and those are allowed to cross the blood brain barrier.

However, it's still much harder to cross the blood brain barrier than it is with regular veins and arteries.

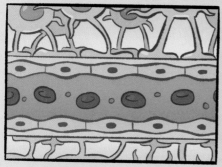

BLOOD BRAIN BARRIER

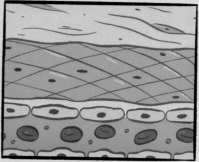

REGULAR VEINS AND ARTERIES

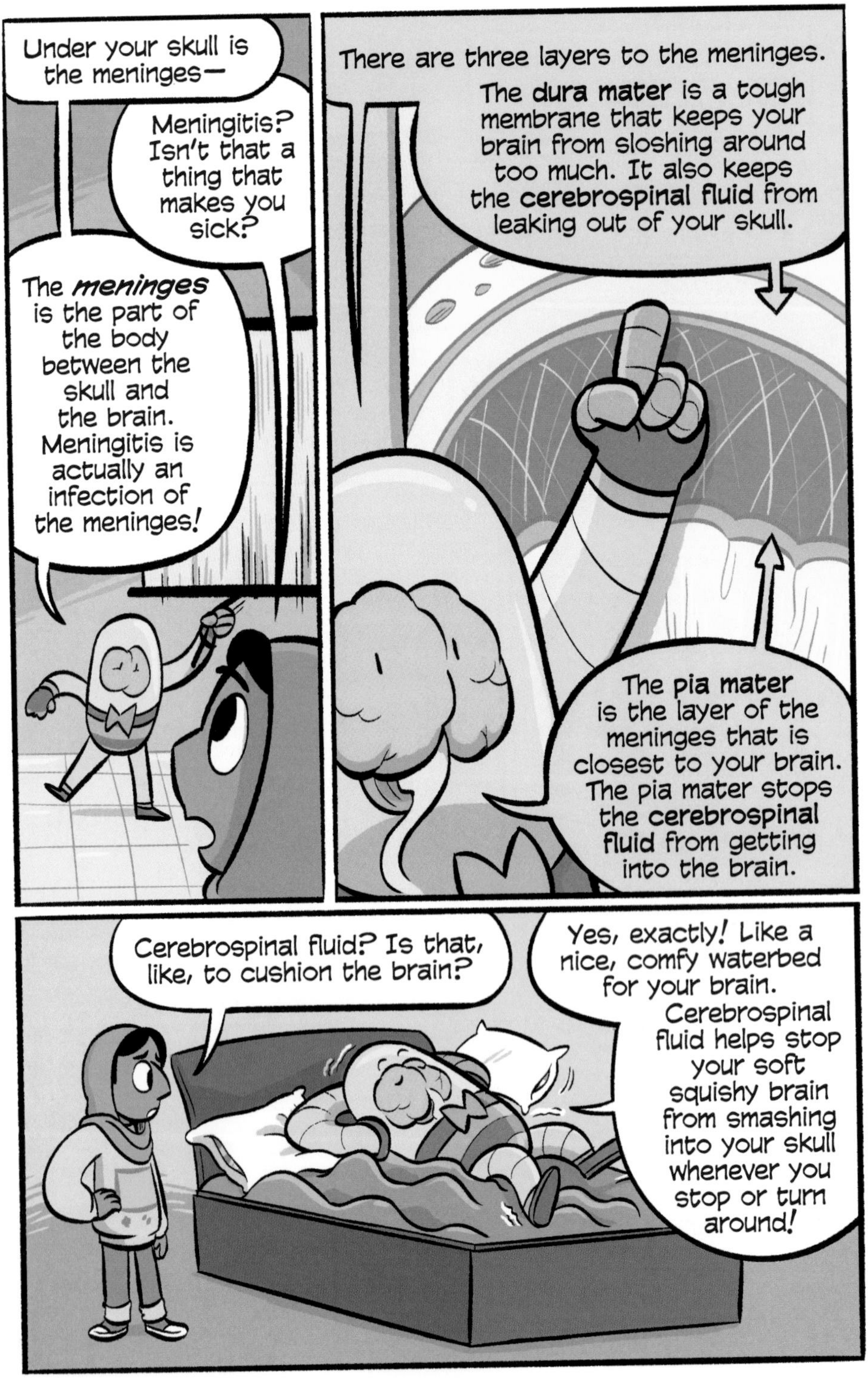

Under your skull is the meninges—
Meningitis? Isn't that a thing that makes you sick?
The *meninges* is the part of the body between the skull and the brain. Meningitis is actually an infection of the meninges!
There are three layers to the meninges.
The **dura mater** is a tough membrane that keeps your brain from sloshing around too much. It also keeps the **cerebrospinal fluid** from leaking out of your skull.
The **pia mater** is the layer of the meninges that is closest to your brain. The pia mater stops the **cerebrospinal fluid** from getting into the brain.
Cerebrospinal fluid? Is that, like, to cushion the brain?
Yes, exactly! Like a nice, comfy waterbed for your brain.
Cerebrospinal fluid helps stop your soft squishy brain from smashing into your skull whenever you stop or turn around!

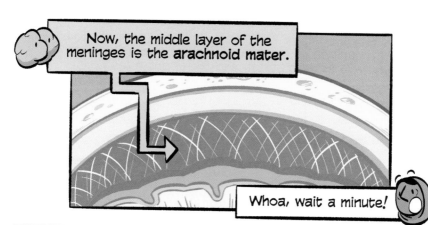

Now, the middle layer of the meninges is the **arachnoid mater**.

Whoa, wait a minute!

"Arachnoid"?
You mean, like a spider?
WE HAVE A SPIDER IN OUR BRAINS?!

Uhhh...

No, it's just called that because it looks a little like a spider's webbing. It helps keep the brain stable.

It's also where the cerebrospinal fluid is, in the arteries—

Oh, okay. That's fine. Better even.

No one wants a spider brain...even if it gave you superpowers.

PAT PAT

So, *uh*, where were we? Oh yes! There are four regions of the brain. Each of these regions have distinct and important roles.

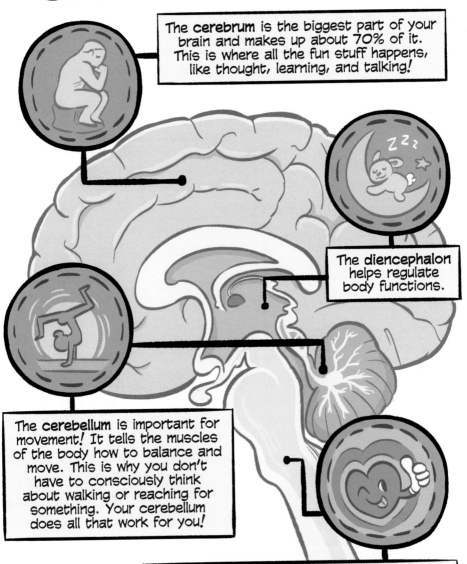

The **cerebrum** is the biggest part of your brain and makes up about 70% of it. This is where all the fun stuff happens, like thought, learning, and talking!

The **diencephalon** helps regulate body functions.

The **cerebellum** is important for movement! It tells the muscles of the body how to balance and move. This is why you don't have to consciously think about walking or reaching for something. Your cerebellum does all that work for you!

The **brain stem** helps run most of your automatic bodily functions. For example, if you're running, the brain stem helps you breathe, and makes your heart beat faster when it needs to. This is the oldest part of your brain, and it was the first to evolve!

The **brain stem** is made up of three main parts:

The **midbrain** regulates vision and hearing.

The **pons** help keep the rhythm of your breathing.

The brain stem regulates things you don't think about, like your heart rate and breathing, but it's also in charge of things like sneezing and barfing!

Oh good.

The **medulla oblongata** is the control center for the heart and lungs!

The brain stem also stimulates arousal, kicking your body into survival mode if it thinks you are in danger. A really good example of this is when you're about to fall asleep...

...and then suddenly you jump back awake for no obvious reason? That actually happens when your brain stem mixes up your body falling asleep for your body falling down!

Sometimes I get confused.

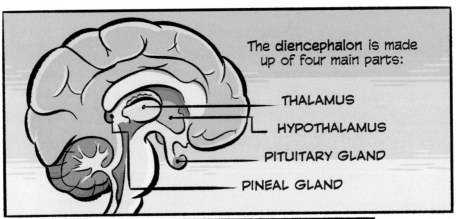

The diencephalon is made up of four main parts:

THALAMUS

HYPOTHALAMUS

PITUITARY GLAND

PINEAL GLAND

The **thalamus** is like the relay center for the brain. It collects and sorts sensory information, and then directs that information to other parts of the body.

The thalamus is also used for things like sleep, language, and learning!

It looks like an egg!

The **hypothalamus** regulates the autonomic nervous system. It tells you when you need to go to sleep or wake up, and whether you're hungry or thirsty. It also regulates your body temperature.

The **pituitary gland** is the "master gland" of the body, is about the size of a pea, and releases hormones that help you grow and develop. The pituitary gland also regulates your blood pressure and metabolism.

The **pineal gland** produces melatonin, which controls your sleep cycle.

The **cerebellum** does three things for you:

The cerebrocerebellum helps the brain plan movements by coordinating movement and sensory information.

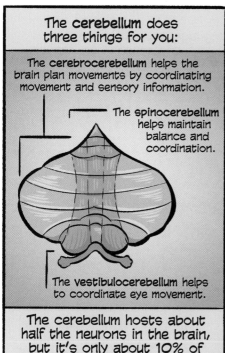

The spinocerebellum helps maintain balance and coordination.

The vestibulocerebellum helps to coordinate eye movement.

The cerebellum hosts about half the neurons in the brain, but it's only about 10% of the total volume of the brain!

The cerebellum is the cutest part of the brain! *Awww,* look at how cute and cuddly it is!

I guess it's cute? In a gooey, *brain-y* sort of way?

The cerebellum helps us maintain our balance and keeps our limbs coordinated. It's also where we store all our learned motions, like martial arts techniques or how to ride a bike.

The cerebellum's job is helping you learn to use your body, but it doesn't create movements on its own. That's where the cerebrum comes in!

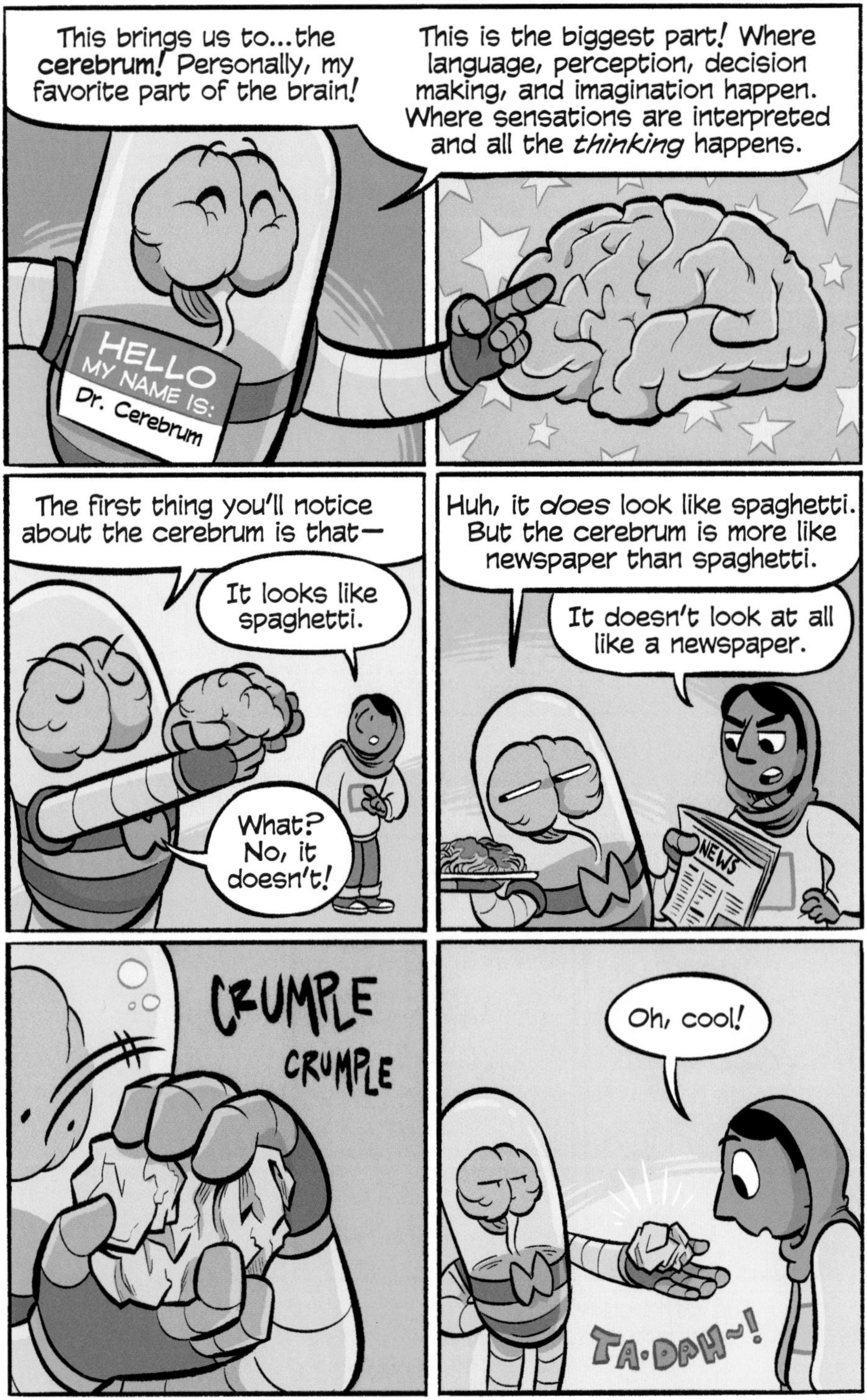
This brings us to...the cerebrum! Personally, my favorite part of the brain!
This is the biggest part! Where language, perception, decision making, and imagination happen. Where sensations are interpreted and all the *thinking* happens.
HELLO MY NAME IS: Dr. Cerebrum
The first thing you'll notice about the cerebrum is that—
It looks like spaghetti.
What? No, it doesn't!
Huh, it *does* look like spaghetti. But the cerebrum is more like newspaper than spaghetti.
It doesn't look at all like a newspaper.
NEWS
CRUMPLE
CRUMPLE
Oh, cool!
TA-DAH~!

The **cerebral cortex** is made up of the cell bodies of neurons. This part of the brain is sometimes called **gray matter** because the cell bodies give it that color.

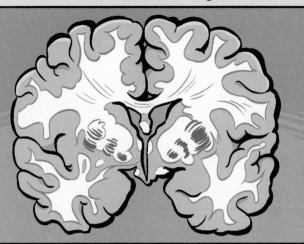

The axons that come off the cell bodies have connections all over the brain. We call this axon-rich layer **white matter,** because the fatty myelin sheaths that cover the axons are white!

The **neocortex** is 90% of the cerebral cortex. The neocortex is like a shell that covers the entire brain. The neocortex comprises 75% of the brain.

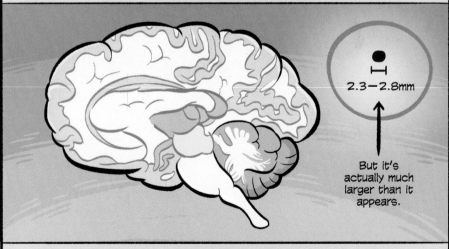

2.3—2.8mm

But it's actually much larger than it appears.

The neocortex is made up of six layers of brain cells on the outermost part of the cerebrum. This thin, outermost covering of the brain is only 2.3—2.8 mm thick, but it's actually really big when you flatten it out!

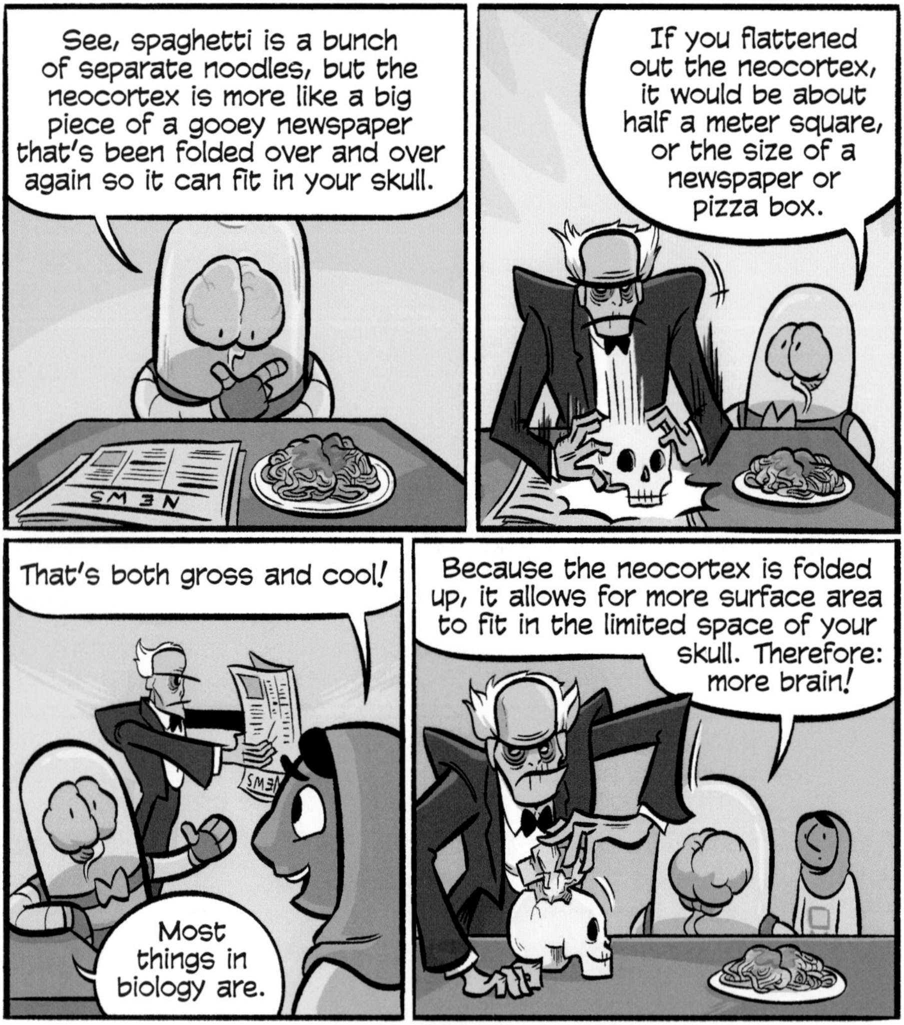
See, spaghetti is a bunch of separate noodles, but the neocortex is more like a big piece of a gooey newspaper that's been folded over and over again so it can fit in your skull.
If you flattened out the neocortex, it would be about half a meter square, or the size of a newspaper or pizza box.
That's both gross and cool!
Most things in biology are.
Because the neocortex is folded up, it allows for more surface area to fit in the limited space of your skull. Therefore: more brain!

The resulting wrinkles in the brain are called the sulcus and the gyrus.

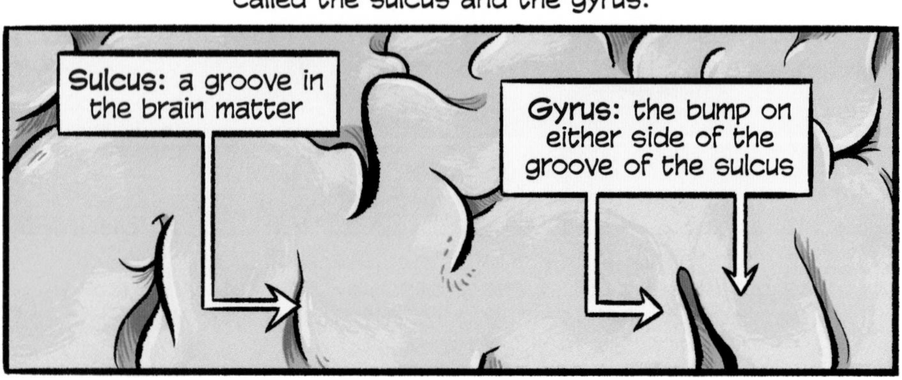
Sulcus: a groove in the brain matter
Gyrus: the bump on either side of the groove of the sulcus

The first thing you'll notice about the cerebrum is—

It looks like pudding?

Yes, it *does* have the consistency of pudding.

Puddingggg...

SQUISH.

The next thing you'll notice after the squishy pudding-y thing is that the brain is divided into two parts—the *left* and *right* hemispheres!

The big valley down the middle of the brain is a special kind of sulcus, called a fissure.

This one is the longitudinal fissure, and it's what separates the two halves of the brain.

All mammals, and bilateral life in general, have brains with two hemispheres.

But why are there two halves of the brain?

Well, the two halves do different things!

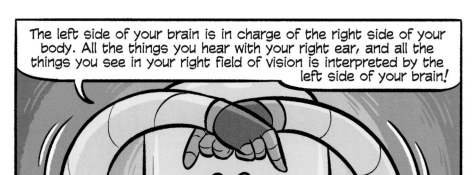

The left side of your brain is in charge of the right side of your body. All the things you hear with your right ear, and all the things you see in your right field of vision is interpreted by the left side of your brain!

The opposite is also true for the right side of your brain. It interprets everything from the left side of your body!

I thought one side of the brain was creative and the other was all about math?

That's not true! Some things like language and speech are laterally focused, but most jobs are handled by both sides of the brain.

The two sides of the brain are different, but they're just two sides of the same coin!

The two hemispheres are connected together by the *corpus callosum*. It's a thick bridge of axons that connect the two hemispheres of the brain. It's like a superhighway that lets the two sides of the brain talk to each other incredibly quickly.

Your cerebrum is divided into four lobes. The lobes are connected to the entire brain and do a lot of different things, but their primary functions are:

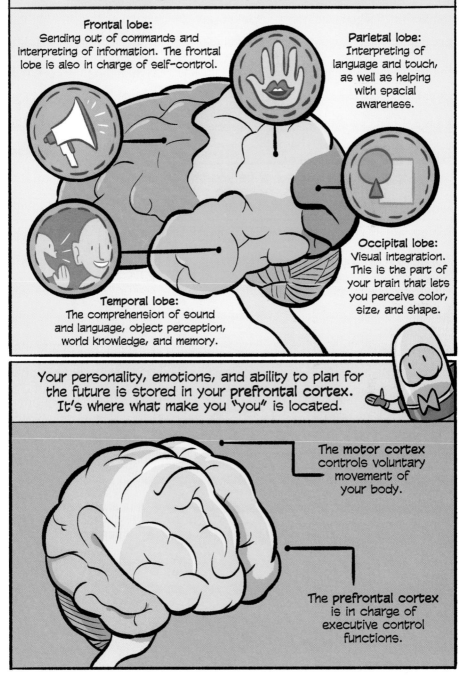

Frontal lobe:
Sending out of commands and interpreting of information. The frontal lobe is also in charge of self-control.

Parietal lobe:
Interpreting of language and touch, as well as helping with spacial awareness.

Occipital lobe:
Visual integration. This is the part of your brain that lets you perceive color, size, and shape.

Temporal lobe:
The comprehension of sound and language, object perception, world knowledge, and memory.

Your personality, emotions, and ability to plan for the future is stored in your **prefrontal cortex**. It's where what make you "you" is located.

The motor cortex controls voluntary movement of your body.

The prefrontal cortex is in charge of executive control functions.

The prefrontal cortex is the most connected area of the brain! This is where you're able to make decisions using previous experience as well as received stimuli.

The prefrontal cortex gives you the ability to make long-term plans, set goals, relate to other people, control your emotions, and even do things like delaying gratification!

There are two kinds of movement: the type you think about, and the type you don't.

I can move without thinking?

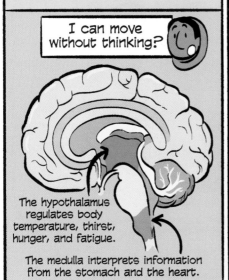

The hypothalamus regulates body temperature, thirst, hunger, and fatigue.

The medulla interprets information from the stomach and the heart.

Of course you do! All the time! Every single beat of your heart, the digestion of food—even moving your hand away from a hot surface—you don't have to think about any of that!

The type of movement that the body does on its own (except for reflexes) is part of the **autonomic nervous system.** Things like breathing, body-temperature regulation, hunger, and thirst are all part of the autonomic nervous system.

All the different parts of your body—your organs, bones, and skin—need to work together for your body to run smoothly.

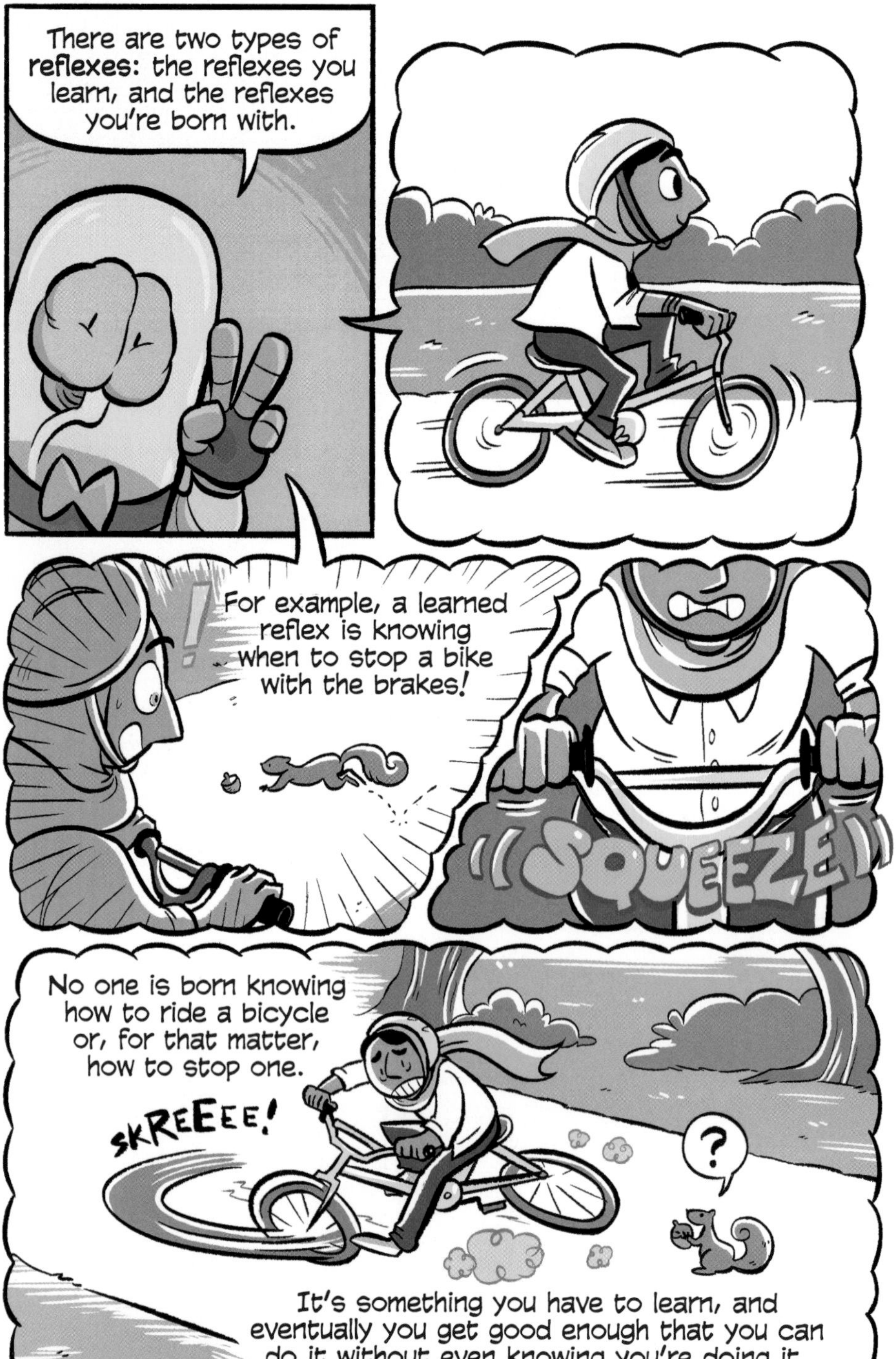
There are two types of **reflexes**: the reflexes you learn, and the reflexes you're born with.
For example, a learned reflex is knowing when to stop a bike with the brakes!
!
SQUEEZE
No one is born knowing how to ride a bicycle or, for that matter, how to stop one.
SKREEEEE!
?
It's something you have to learn, and eventually you get good enough that you can do it without even knowing you're doing it.

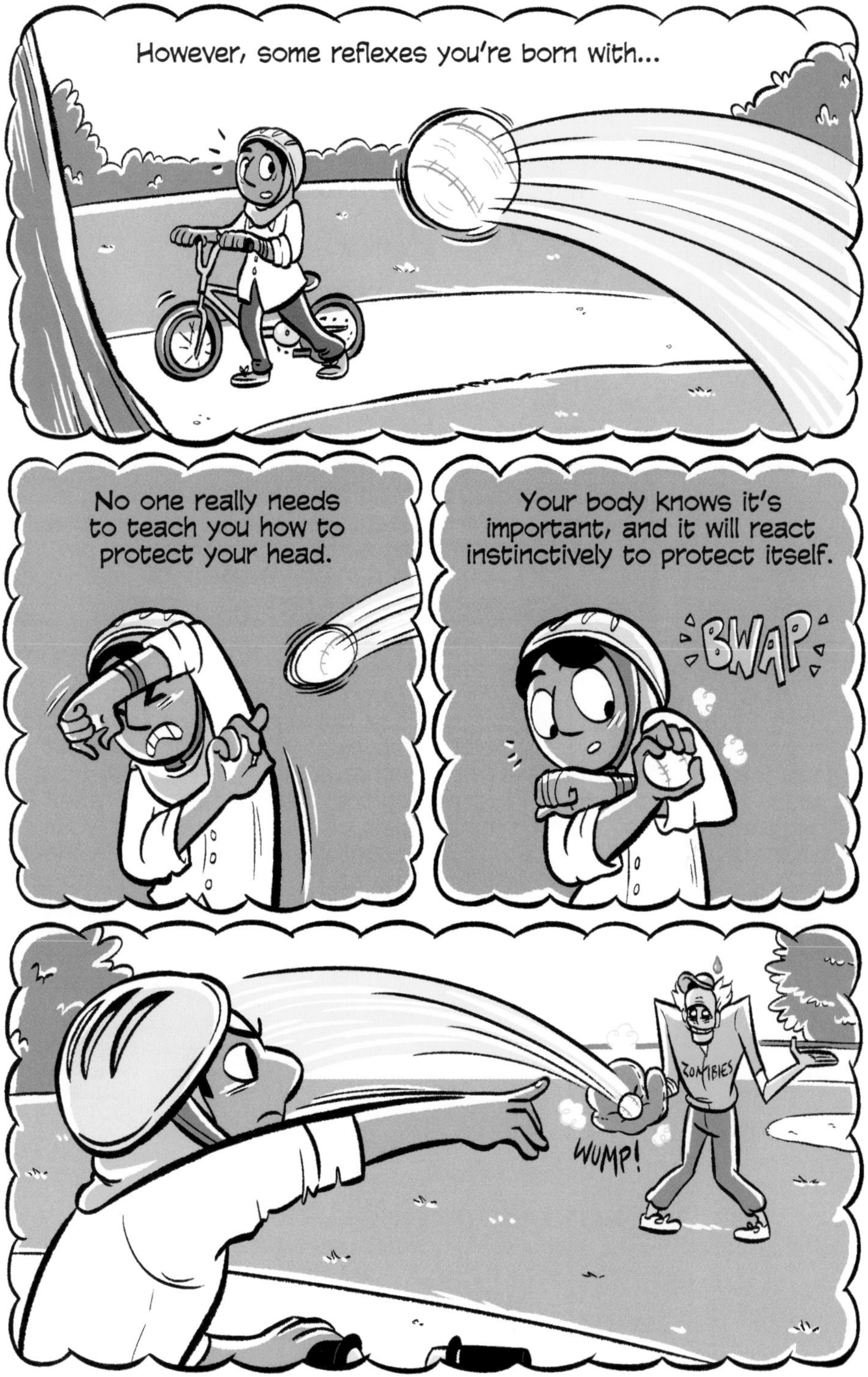
However, some reflexes you're born with...
No one really needs to teach you how to protect your head.
Your body knows it's important, and it will react instinctively to protect itself.
BWAP
ZOMBIES
WUMP!

There are also withdrawal reflexes!

HEY! Ow!!

When your body receives a stimulus...

(in this case, my kick)

...the sensory receptors detects the stimulus and nerves then send that information up the peripheral nervous system, to the spine and central nervous system.

The action potential reaches the integration center in your spine, and that information is passed on to two different places: part of it goes to the brain...

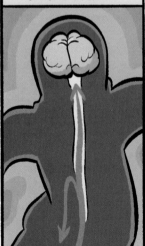

...but motor neurons will also send directions back down the spine to the peripheral nervous system, to the original point of stimulus—causing you to react and move your leg away!

Finally, your brain will get the memo, and it will figure out what caused the stimulus. A reflex is an action your body takes without you telling it to do so. This lets you react faster and stay alive longer!

Why am I always the last to know?

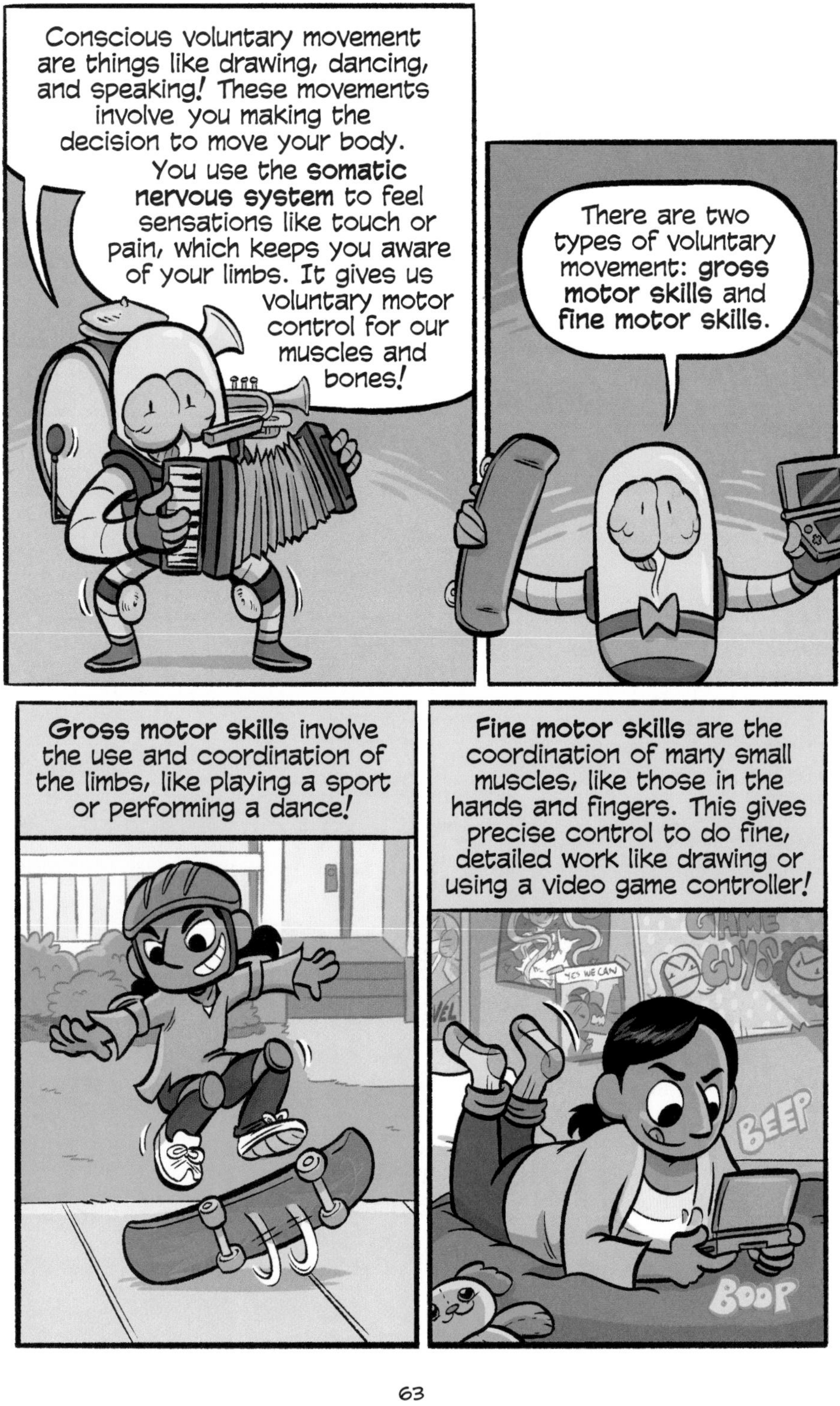
Conscious voluntary movement are things like drawing, dancing, and speaking! These movements involve you making the decision to move your body.
You use the **somatic nervous system** to feel sensations like touch or pain, which keeps you aware of your limbs. It gives us voluntary motor control for our muscles and bones!
There are two types of voluntary movement: **gross motor skills** and **fine motor skills**.
Gross motor skills involve the use and coordination of the limbs, like playing a sport or performing a dance!
Fine motor skills are the coordination of many small muscles, like those in the hands and fingers. This gives precise control to do fine, detailed work like drawing or using a video game controller!
GAME GUYS
YES WE CAN
BEEP
BOOP

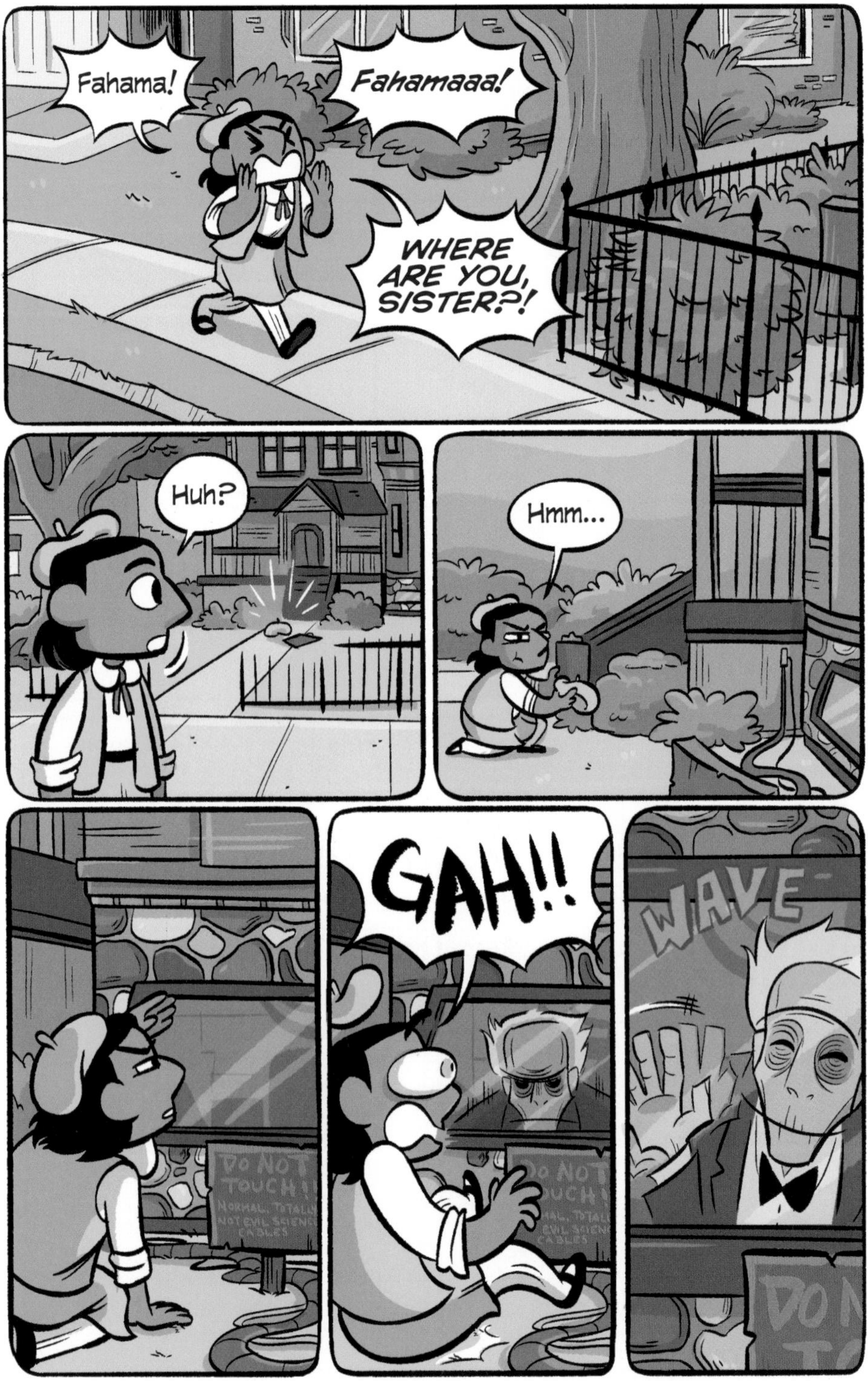
Fahama!
Fahamaaa!
WHERE ARE YOU, SISTER?!
Huh?
Hmm...
DO NOT TOUCH!!
NORMAL, TOTALLY NOT EVIL SCIENC
CABLES
GAH!!
WAVE

The **senses** are how a living thing is able to gather information about the world around it.

Our five senses (touch, taste, smell, sight, and hearing) receive and react to three types of stimuli.

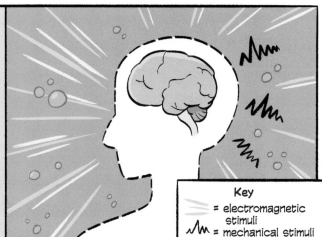

Key

= electromagnetic stimuli

= mechanical stimuli

= chemical stimuli

The senses do this by **transduction**.

That's the transformation of physical information into action potentials the brain can understand.

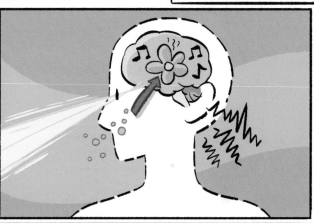

A lot of the brain is dedicated to interpreting and understanding what our senses tell us.

That's what **perception** is: the organization, interpretation, and contextualization of sensory stimuli.

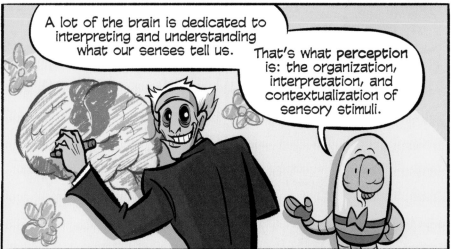

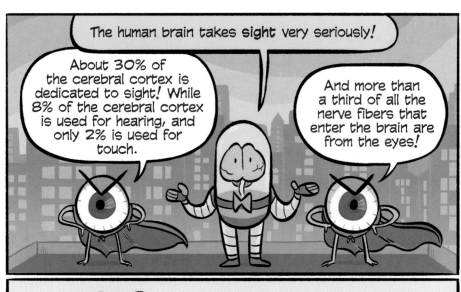

The human brain takes **sight** very seriously!

About 30% of the cerebral cortex is dedicated to sight! While 8% of the cerebral cortex is used for hearing, and only 2% is used for touch.

And more than a third of all the nerve fibers that enter the brain are from the eyes!

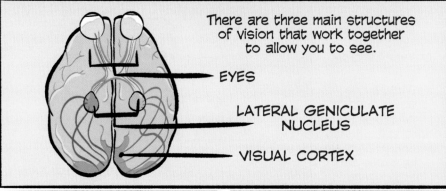

There are three main structures of vision that work together to allow you to see.

EYES

LATERAL GENICULATE NUCLEUS

VISUAL CORTEX

But before I teach you *how* we see, I need to teach you *what* we see. Tell me, what *do* you see?

A teapot?

Right. But what you're actually seeing is the *reflection* of light off the teapot!

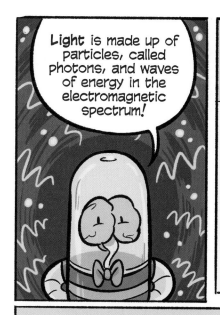

Light is made up of particles, called photons, and waves of energy in the electromagnetic spectrum!

Light waves are measured from the top of one wave to the next and in units called nanometers (nm). This measurement is one billionth of a meter.

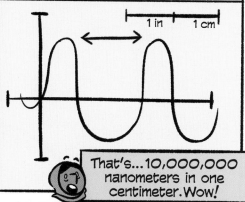

1 in   1 cm

That's...10,000,000 nanometers in one centimeter. Wow!

The colors that we can see are only a tiny fraction of the available electromagnetic spectrum.

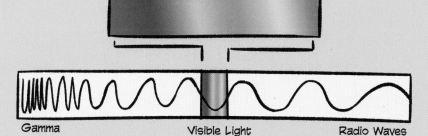

Gamma   Visible Light   Radio Waves

The wavelengths of light that the human eye can perceive is between 390 to 700 nm.

Light that has a short wavelength will result in light that will appear to be blue.

A longer wavelength will mean the light will be perceived as red.

450–495 nm   620–750 nm

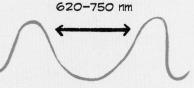

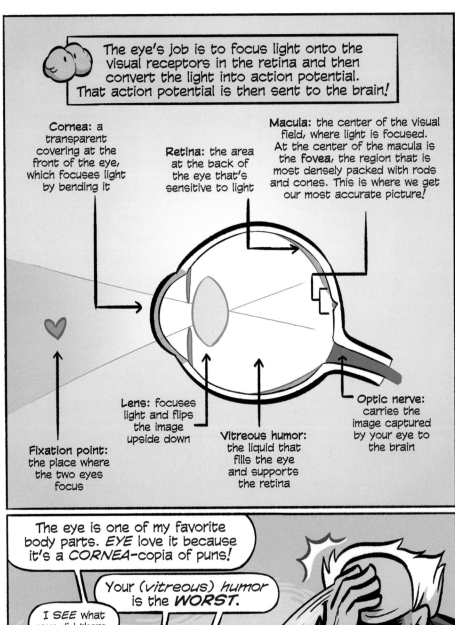

The eye's job is to focus light onto the visual receptors in the retina and then convert the light into action potential. That action potential is then sent to the brain!

Cornea: a transparent covering at the front of the eye, which focuses light by bending it

Retina: the area at the back of the eye that's sensitive to light

Macula: the center of the visual field, where light is focused. At the center of the macula is the **fovea**, the region that is most densely packed with rods and cones. This is where we get our most accurate picture!

Lens: focuses light and flips the image upside down

Vitreous humor: the liquid that fills the eye and supports the retina

Optic nerve: carries the image captured by your eye to the brain

Fixation point: the place where the two eyes focus

The eye is one of my favorite body parts. *EYE* love it because it's a *CORNEA*-copia of puns!

Your (*vitreous*) *humor* is the **WORST.**

I *SEE* what you did there.

What can I say? I'm a good *PUPIL.*

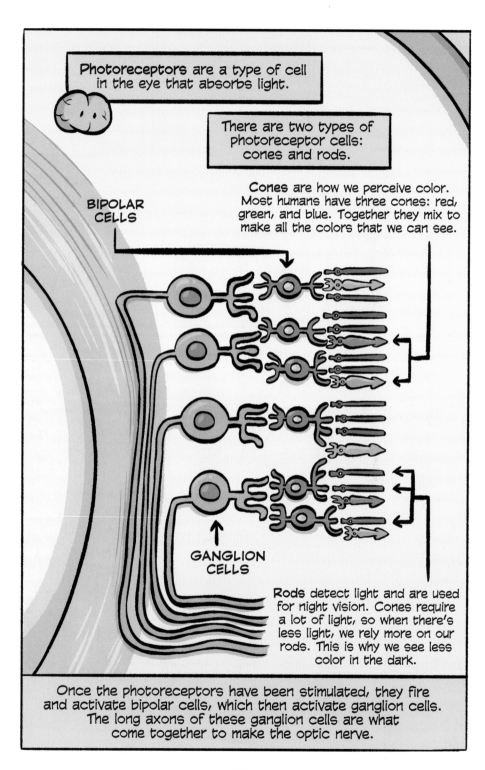

Photoreceptors are a type of cell in the eye that absorbs light.

There are two types of photoreceptor cells: cones and rods.

Cones are how we perceive color. Most humans have three cones: red, green, and blue. Together they mix to make all the colors that we can see.

BIPOLAR CELLS

GANGLION CELLS

Rods detect light and are used for night vision. Cones require a lot of light, so when there's less light, we rely more on our rods. This is why we see less color in the dark.

Once the photoreceptors have been stimulated, they fire and activate bipolar cells, which then activate ganglion cells. The long axons of these ganglion cells are what come together to make the optic nerve.

We're the cones in your eye! We let you perceive color.

Yeah!

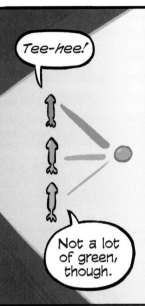

Tee-hee!

Not a lot of green, though.

Rods and cones are what translate the light that goes into your eyes into action potentials the brain can understand.

Information in the form of action potential

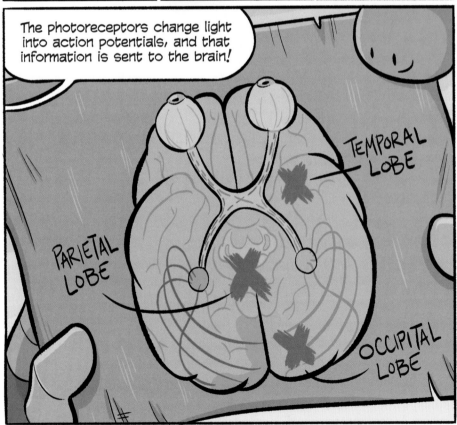

The photoreceptors change light into action potentials, and that information is sent to the brain!

TEMPORAL LOBE

PARIETAL LOBE

OCCIPITAL LOBE

70

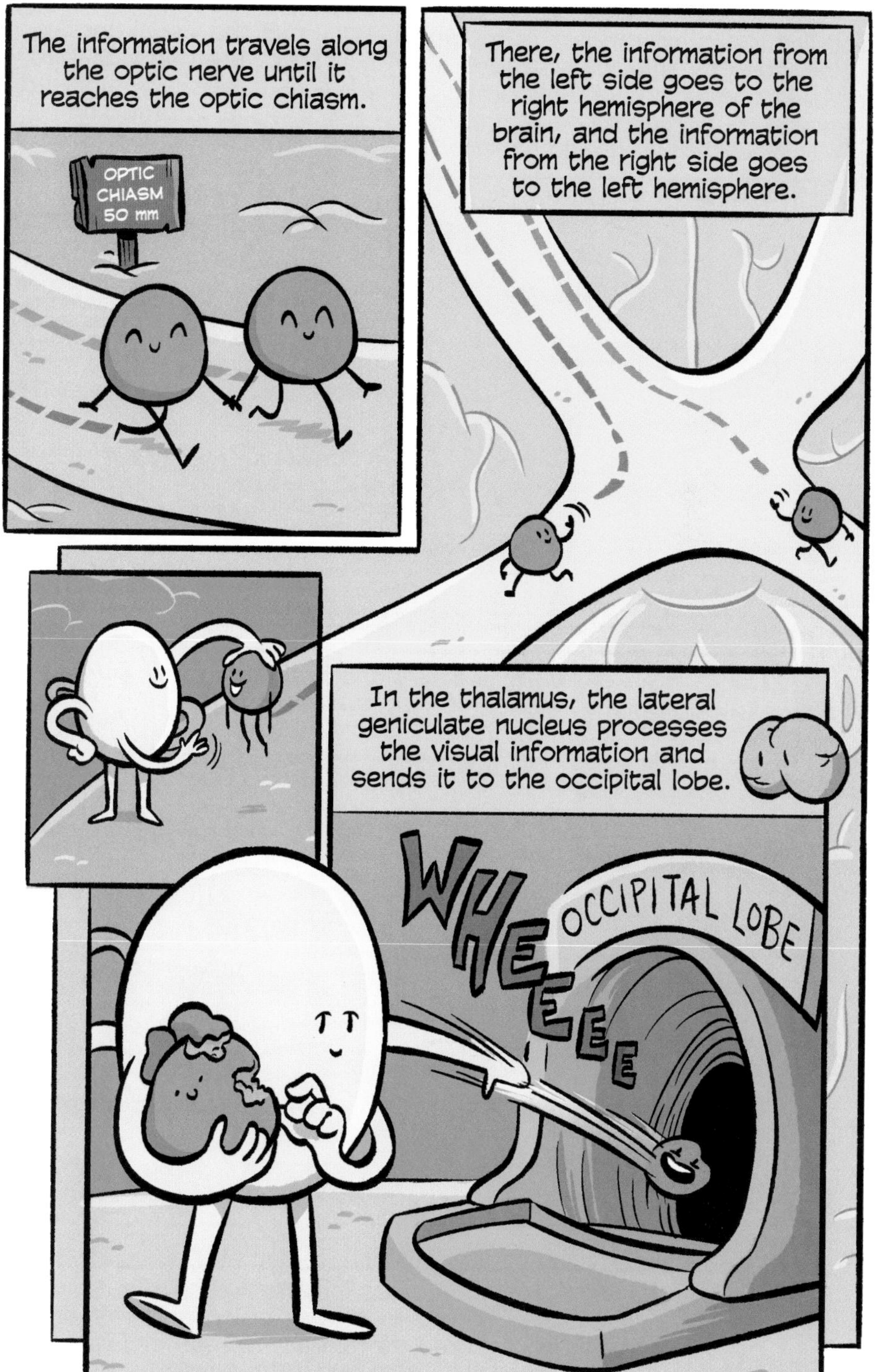

The information travels along the optic nerve until it reaches the optic chiasm.
OPTIC CHIASM 50 mm
There, the information from the left side goes to the right hemisphere of the brain, and the information from the right side goes to the left hemisphere.
In the thalamus, the lateral geniculate nucleus processes the visual information and sends it to the occipital lobe.
WHEEEEE
OCCIPITAL LOBE

The visual information is roughly decoded by the occipital lobe and visual cortex and is sent on to the cerebral cortex to be integrated with your memories and experience.

WELCOME to the OCCIPITAL LOBE

From the visual cortex, there are two highways that take the processed visual information away to be analyzed: the dorsal stream and the ventral stream.

DORSAL STREAM

WOO! HOO!

VENTRAL STREAM

The *dorsal stream* is where the brain figures out whether something is moving, where it is, how big it is, and how fast it's moving.

The *ventral stream* takes information to the temporal lobe, where it is sorted by shape, color, and patterns, until it is recognized as familiar objects or ideas. This is where we figure out what something is!

Okay, let's move on to *touch*, shall we?

That's easy. It's what you can feel with your hands.

It's more than that! Touch is the border between what is "you" and what is "not you."

There are two types of touch: passive and active.

Passive touch is when something unexpectedly touches your skin.

TOSS!

BONK!

You recognize that you've been touched before you know what has actually touched you.

Active touch, on the other hand...

(Get it? The other hand!)

Ugh.

When we touch something on purpose, this is active touch. In this case, you're usually using your hands in order to touch something.

Now, mechanoreceptors are specialized ends of unipolar neurons in your glabrous skin.

What's—?

Glabrous skin is any part of your body that doesn't have hair.

Oh.

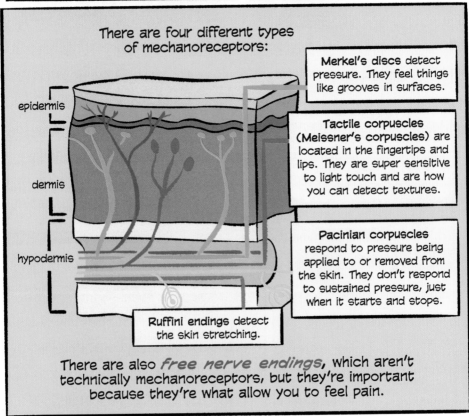

There are four different types of mechanoreceptors:

epidermis

dermis

hypodermis

Merkel's discs detect pressure. They feel things like grooves in surfaces.

Tactile corpuscles (Meissner's corpuscles) are located in the fingertips and lips. They are super sensitive to light touch and are how you can detect textures.

Pacinian corpuscles respond to pressure being applied to or removed from the skin. They don't respond to sustained pressure, just when it starts and stops.

Ruffini endings detect the skin stretching.

There are also *free nerve endings*, which aren't technically mechanoreceptors, but they're important because they're what allow you to feel pain.

Merkel's discs, Ruffini endings, and free nerve endings are known as "slow-adapting" receptors, because these axons will continue to fire until the stimulus is removed.

Here's an example:

Hey, there's a cold thing touching us!

Hunh? Really?

It's cold! It's cold!

Hmm, a cold thing you say?

IT'S SO COLD!!!

I'll put the ice cube in the glass.

Pacinian corpuscles only fire with new information, when there's a change.

Hey, did you know you have pants on?

Yes, I do!

The **parietal lobe** is used to interpret sensory information and perception. It integrates all the information gathered from our senses and helps us understand our surroundings, allowing us to coordinate our movements.

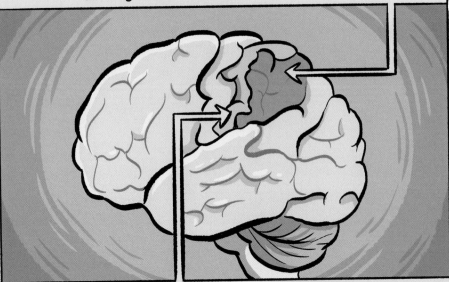

The **somatosensory cortex** processes information taken from your skin, things like pressure and temperature.

Stimulus from the left side of the body will activate the right side of your primary somatosensory cortex, and vice versa!

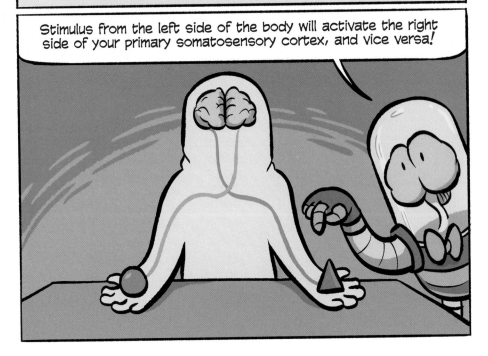

When touch receptors are stimulated, that information is sent to the brain and processed in the somatosensory cortex.

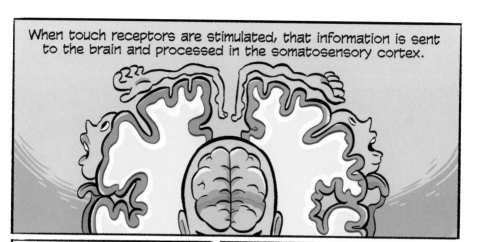

When you use only touch to figure out what something is, you're using your *haptic perception*.

This is your ability to determine details, or even identify an object by touching it and feeling its surface, as opposed to having it passively come in contact with your skin.

This is a pencil, and this is a pen!

Correct! You just used your haptic perception.

Okay, am I touching you now with the pen or the pencil?

I-I'm not sure!

That's because elbows weren't designed to give you the same level of precision as hands!

And the nose, the mountain of the face! The crowning jewel of the head!

Smell and *taste* are senses that interpret chemicals rather than energy, distinguishing them from the other senses.

We're going to be talking a lot about snot, aren't we?

What? No! Never. I'm offended you would say such a thing.

Mucus, yes. Snot, no.

Mucus is snot.

Well, it's a very special snot called **olfactory mucus.** It helps you smell!

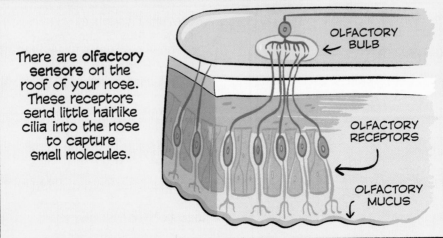

There are **olfactory sensors** on the roof of your nose. These receptors send little hairlike cilia into the nose to capture smell molecules.

OLFACTORY BULB

OLFACTORY RECEPTORS

OLFACTORY MUCUS

In your nose, there are 10 million olfactory receptor cilia, and there are roughly 350 types of them!

Each olfactory receptor responds to different types of molecules. Once the receptor is stimulated, it transmits an action potential.

The action potential then travels up the stimulated olfactory receptor until it reaches the olfactory bulb.

The olfactory bulb sends its information to several different areas of the cerebral cortex, each of which handles different aspects of smelling.

The pyriform cortex helps you recognize smells on an unconscious level.

The orbitofrontal cortex is where the axons of the olfactory sensory neurons meet with the dendrites of the brain and form little bundles, and it's where you consciously recognize smell.

The entorhinal cortex ties smell to your memories and hormones.

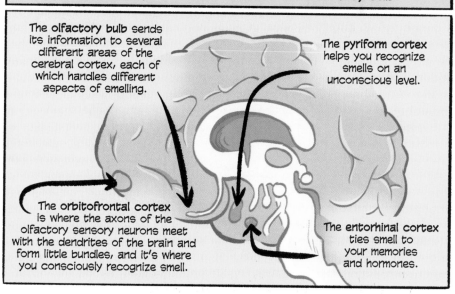

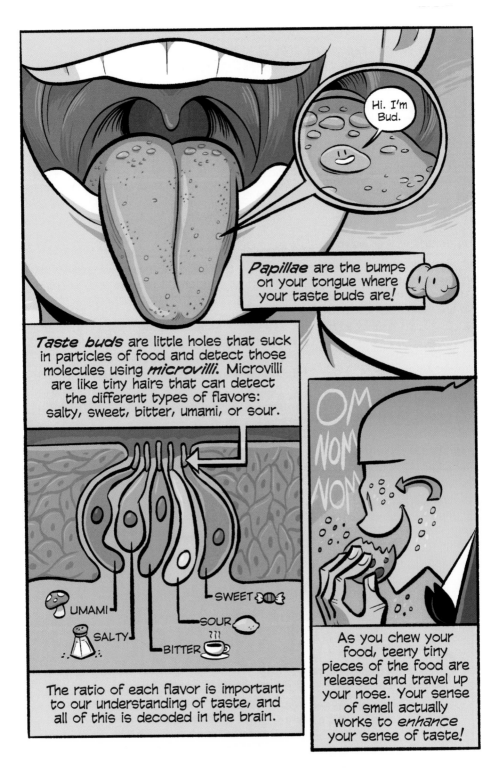

Hi. I'm Bud.

Papillae are the bumps on your tongue where your taste buds are!

Taste buds are little holes that suck in particles of food and detect those molecules using *microvilli*. Microvilli are like tiny hairs that can detect the different types of flavors: salty, sweet, bitter, umami, or sour.

UMAMI

SALTY

BITTER

SOUR

SWEET

The ratio of each flavor is important to our understanding of taste, and all of this is decoded in the brain.

OM NOM NOM

As you chew your food, teeny tiny pieces of the food are released and travel up your nose. Your sense of smell actually works to *enhance* your sense of taste!

What we perceive as "sound" is our brain interpreting vibrations in the air around us. These vibrations are called **sound waves**.

That *sounds* really simple—

Tee-hee-hee.

HONK!

WHHHSSH

BLAH.

BLAH. BLAH.

!

But seriously, though, sound gives us a lot of information that we often don't consciously realize! Pitch, loudness, direction, even language, are just a few examples.

CLICK!

When the ear receives a stimulus, it's your brain that changes it into hearing.

Sound is created when something vibrates! A vibration, or *frequency*, is measured in hertz—the number of vibrations per second. Human hearing spans 20–20,000 hertz. We interpret frequency as pitch.

The height of a sound wave determines the loudness of a sound. This is measured in decibels and is all about how hard the wave hits your ear.

The ear is a complex organ and actually has more in common with touch than it does with sight because it's a mechanical sense!

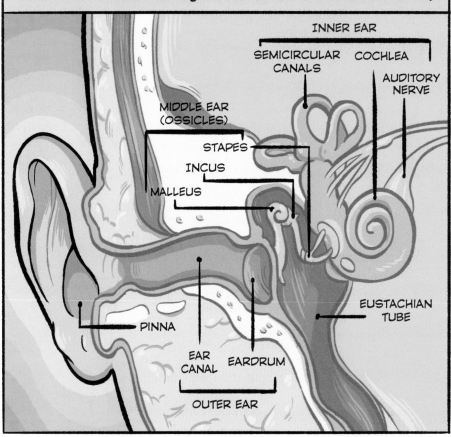

INNER EAR

SEMICIRCULAR CANALS

COCHLEA

AUDITORY NERVE

MIDDLE EAR (OSSICLES)

STAPES

INCUS

MALLEUS

EUSTACHIAN TUBE

PINNA

EAR CANAL

EARDRUM

OUTER EAR

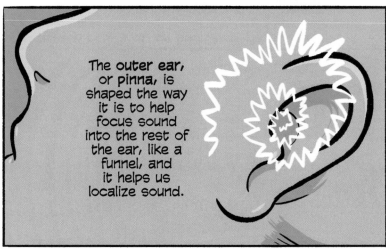

The outer ear, or pinna, is shaped the way it is to help focus sound into the rest of the ear, like a funnel, and it helps us localize sound.

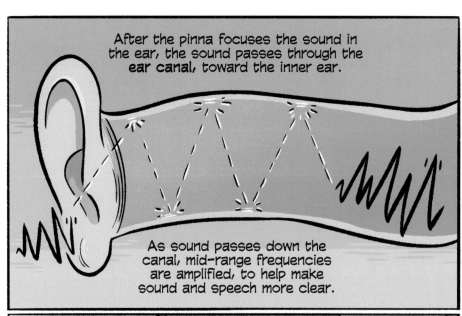

After the pinna focuses the sound in the ear, the sound passes through the **ear canal,** toward the inner ear.

As sound passes down the canal, mid-range frequencies are amplified, to help make sound and speech more clear.

The ear canal also protects the eardrum, which is delicate, you guys!

After the sound has been collected by the pinna and amplified by the auditory canal, it finally reaches the **eardrum** (also known as the tympanic membrane), causing these amplified vibrations to move into the middle ear.

WHAM

The eardrum converts the sound wave into physical vibrations that travel along the **ossicles,** a chain of three tiny bones in the middle ear.

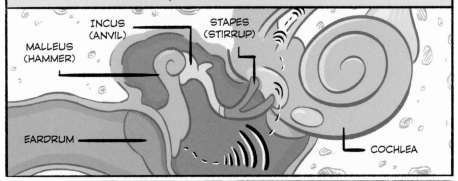

INCUS
(ANVIL)

STAPES
(STIRRUP)

MALLEUS
(HAMMER)

EARDRUM

COCHLEA

The vibrations from the ossicles are transmitted to the cochlea by vibrating a small membrane at the front of the cochlea called the oval window.

PUSH!

OVAL WINDOW

SPLOOSH

COCHLEAR FLUID

The cochlea is a spiral cavity filled with fluid. It's a little like a seashell deep in the inner ear. The inside of the cochlea is covered in little hairlike cells known as **stereocilia.**

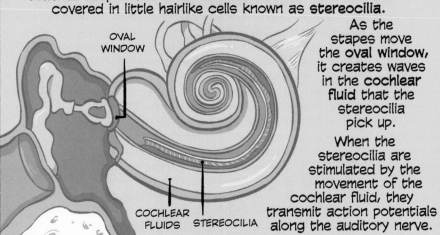

OVAL
WINDOW

COCHLEAR
FLUIDS

STEREOCILIA

As the stapes move the **oval window,** it creates waves in the **cochlear fluid** that the stereocilia pick up.

When the stereocilia are stimulated by the movement of the cochlear fluid, they transmit action potentials along the auditory nerve.

The movement inside the cochlea stimulates nerve impulses or action potentials that then pass along the auditory nerve.

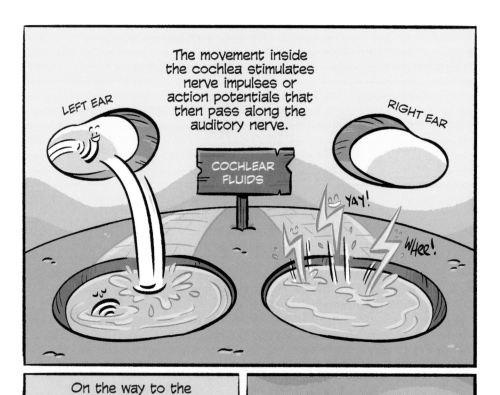

LEFT EAR

RIGHT EAR

COCHLEAR FLUIDS

YAY!

WHEE!

On the way to the auditory cortex, the sound first switches from the left ear to the right side of the brain, and from the right ear to the left side of the brain.

AUDITORY NERVE

The thalamus relays the sound information to the auditory cortex.

THALAMUS

The temporal lobe of the brain helps interpret complex sounds like language and music, as well as the basic units of sound like pitch and frequency.

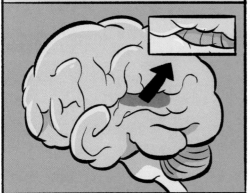

In your auditory cortex, there is a map called the tonotopic map. However, it doesn't map locations—it maps auditory frequencies.

The tonotopic map decodes the information from the cochlea, and this is how we tell the difference between the sound of a car engine and your favorite song.

Hearing is a team sport! Your brain can figure out where a sound is coming from by comparing the delay between the two ears receiving a particular sound.

Let's talk about *language.*

Okay, *let's.*

A baby crying and a dog wagging its tail are both forms of communication.

Your face can also communicate a great deal without you even saying a word.

Your facial expressions can also add meaning to what you say!

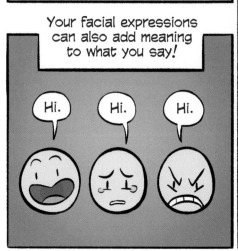

Hi.

Hi.

Hi.

Communication shapes our world more than we realize, and language is simply a more complex way to communicate.

In 1861, Paul Broca worked with a patient who had lost the ability to form speech, even though the patient could still understand what was being said to him.

After examining the patient's brain, Broca discovered what is now known as **Broca's area**.

Wernicke's area was discovered by Carl Wernicke around 1874. He had several patients who could speak, but only gibberish, and couldn't understand anything being said to them.

Wernicke's and Broca's areas are connected by the **arcuate fasciculus**. It's like a superhighway between the two areas that allows them to communicate with each other very quickly.

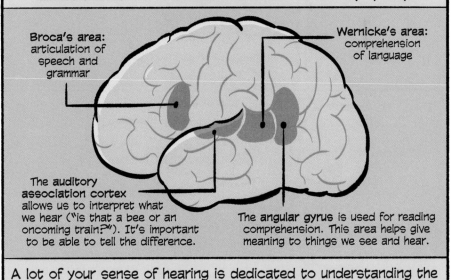

Broca's area: articulation of speech and grammar

Wernicke's area: comprehension of language

The **auditory association cortex** allows us to interpret what we hear ("is that a bee or an oncoming train?"). It's important to be able to tell the difference.

The **angular gyrus** is used for reading comprehension. This area helps give meaning to things we see and hear.

A lot of your sense of hearing is dedicated to understanding the spoken word, but being able to create speech is important too.

Vocalizing is surprisingly complex, considering how often we use it without even thinking.

You have to control your breathing, the muscles of your larynx and vocal chords, as well as your lips and tongue.

LIPS →

TONGUE

LARYNX

VOCAL CORDS

Ma-ma!

Da-da!

The basal ganglia and the thalamus are the major regions of the brain that work with the frontal lobe to learn language.

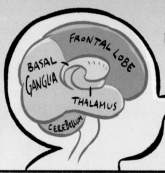

FRONTAL LOBE

BASAL GANGLIA

THALAMUS

CEREBELLUM

When you're learning to speak, the cerebellum is very active because you're learning to coordinate parts of your body.

Hello, Father,

I regret to inform you that I have soiled myself.

Once you have a handle on language, the cerebellum takes a back seat and lets the rest of your brain do the heavy lifting.

90

As I'm sure you've noticed by now, brains are elaborate things!

Whrrrr

Peeoo

Well, memory is just as complex!

The three main types of memory, **sensory memory**, **working (short-term) memory**, and **long-term memory**, all work together to allow you to access, use, and create memories.

SHORT-TERM MEMORY

SENSORY MEMORY

LONG-TERM MEMORY

What you probably think of as "memory" is the transfer of information from your short-term memory to your long-term memory.

On the surface, memory seems incredibly straightforward—you either remember something or you don't.

But if you dig a little deeper, you'll see that this is very far from the truth!

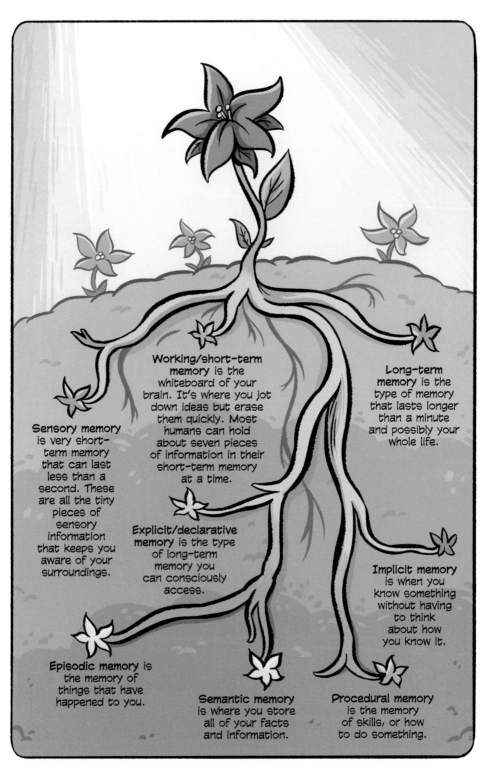

Working/short-term memory is the whiteboard of your brain. It's where you jot down ideas but erase them quickly. Most humans can hold about seven pieces of information in their short-term memory at a time.

Long-term memory is the type of memory that lasts longer than a minute and possibly your whole life.

Sensory memory is very short-term memory that can last less than a second. These are all the tiny pieces of sensory information that keeps you aware of your surroundings.

Explicit/declarative memory is the type of long-term memory you can consciously access.

Implicit memory is when you know something without having to think about how you know it.

Episodic memory is the memory of things that have happened to you.

Semantic memory is where you store all of your facts and information.

Procedural memory is the memory of skills, or how to do something.

FATHER!

Gah!!

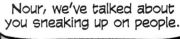

Nour, we've talked about you sneaking up on people.

I cannot help that I have the natural skills of a ninja.

≷Sigh≷

Well, I guess I'm going to need more groceries now. Go to the shop and get me some more milk, eggs, and sugar.

I suspect Fahama has been abducted by a well-dressed zombie. Should I deal with that first?

Go to the shop first, and then you can go back to your zombie dress-up game.

Affirmative!

So when you're given a list of things to remember, you store that in your **working**, or **short-term**, memory, which is overseen by your frontal lobe.

You remember how to ride a bike, but you don't have to think about it because the skill is saved in your cerebellum as a **procedural memory**.

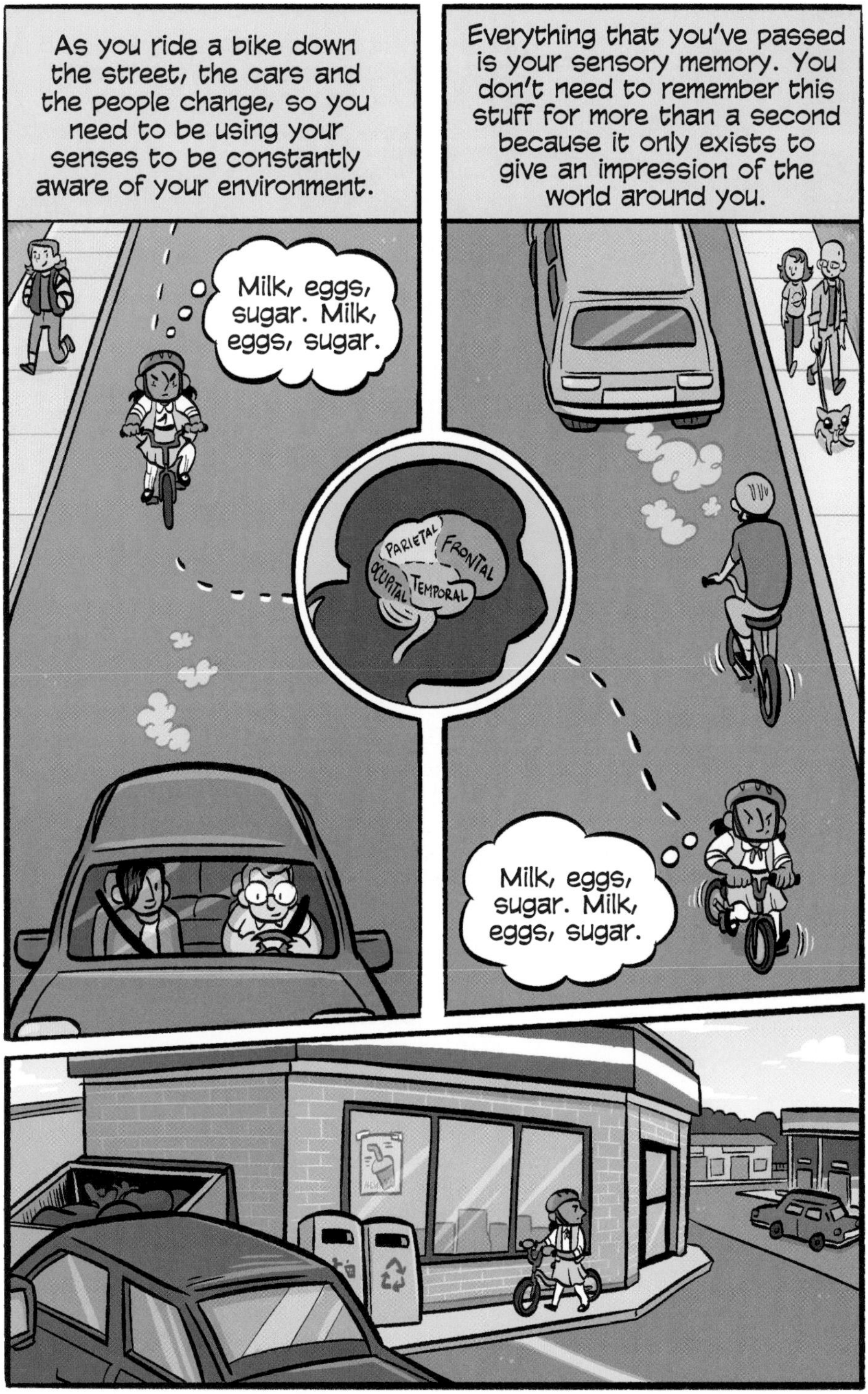
As you ride a bike down the street, the cars and the people change, so you need to be using your senses to be constantly aware of your environment.
Milk, eggs, sugar. Milk, eggs, sugar.
Everything that you've passed is your sensory memory. You don't need to remember this stuff for more than a second because it only exists to give an impression of the world around you.
PARIETAL
FRONTAL
OCCIPITAL
TEMPORAL
Milk, eggs, sugar. Milk, eggs, sugar.

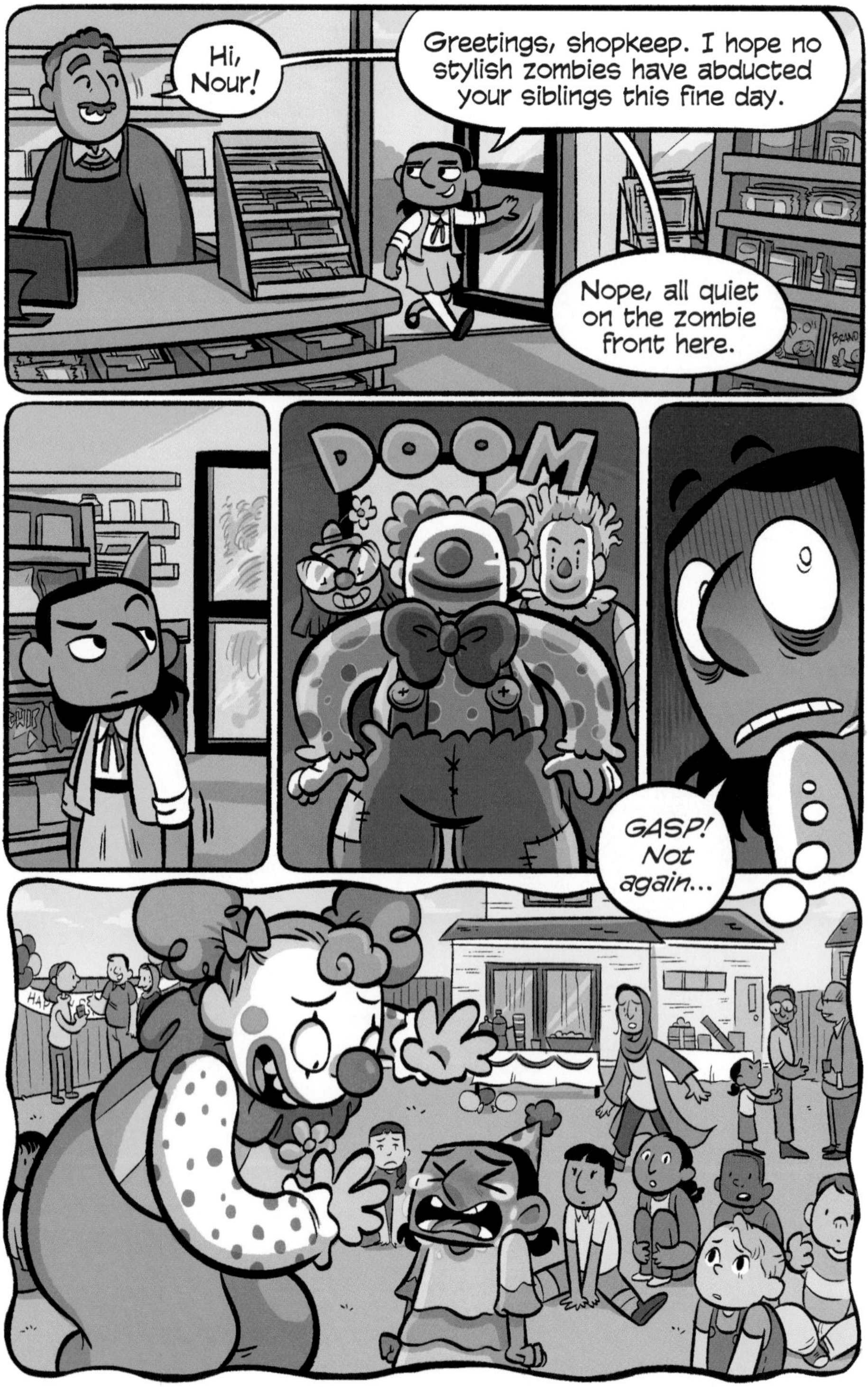
Hi, Nour!
Greetings, shopkeep. I hope no stylish zombies have abducted your siblings this fine day.
Nope, all quiet on the zombie front here.
DOOM
GASP! Not again...

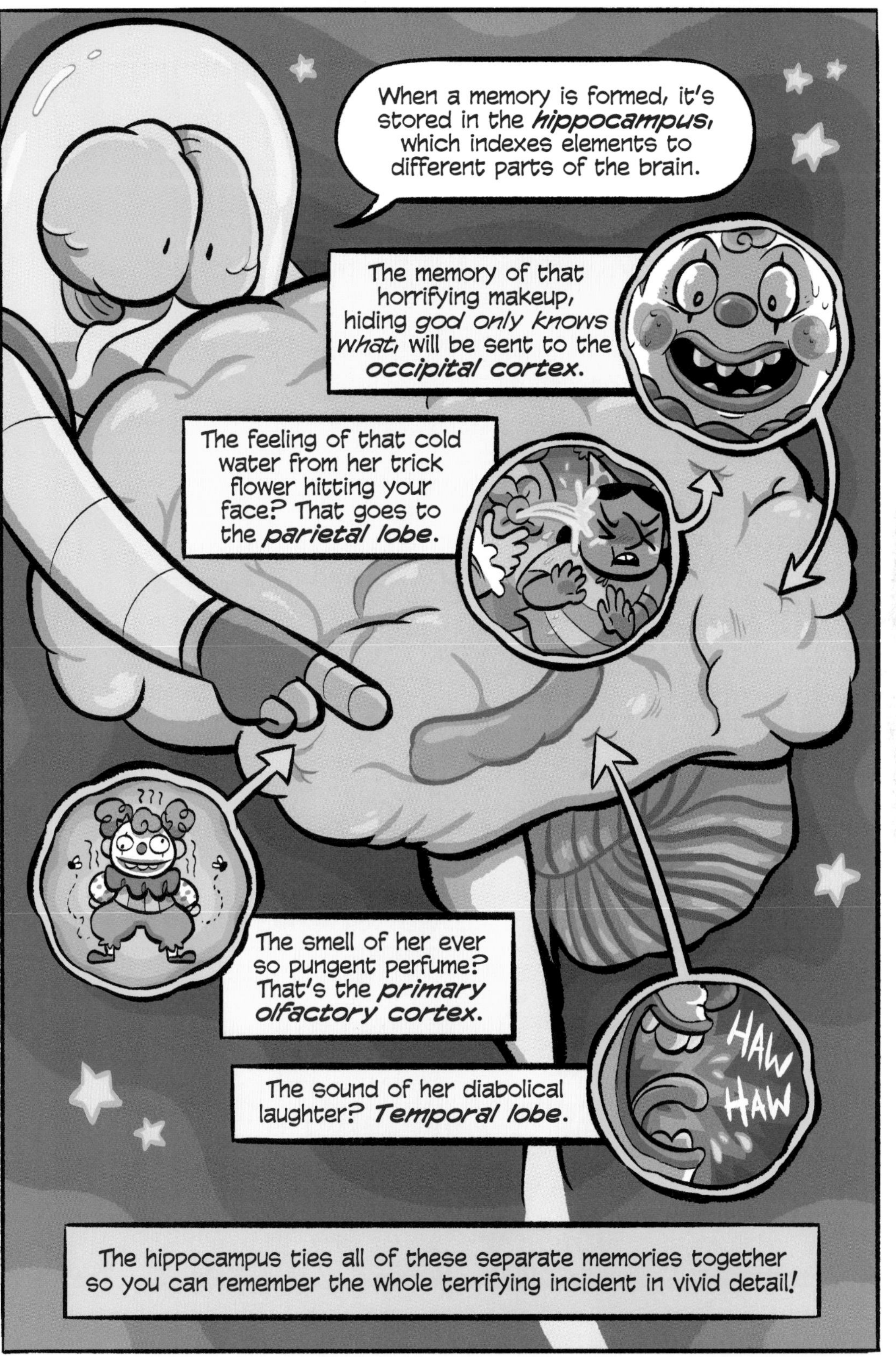

When a memory is formed, it's stored in the *hippocampus*, which indexes elements to different parts of the brain.
The memory of that horrifying makeup, hiding *god only knows what*, will be sent to the ***occipital cortex***.
The feeling of that cold water from her trick flower hitting your face? That goes to the *parietal lobe*.
???
The smell of her ever so pungent perfume? That's the *primary olfactory cortex*.
The sound of her diabolical laughter? *Temporal lobe*.
HAW HAW
The hippocampus ties all of these separate memories together so you can remember the whole terrifying incident in vivid detail!

Semantic memory is the other type of explicit, or declarative, memory. This is the type of memory where you can remember facts, figures, and information without necessarily knowing where and when you learned it.

There are a few reasons why you forget things. Either the memory failed to encode, which means the memory never passed from your working memory to your long-term memory, or sometimes the memory is actually just lost (although this is rare)!

SHORT-TERM MEMORY   LONG-TERM MEMORY

Or you might not have actually forgotten, you might have just failed to retrieve the memory. So the memory could still be there, but you're having trouble getting at it.

Your short-term memory can only hold a limited amount of information for a period of time.

If something happens that takes priority in your short-term working memory—like a clown crisis, for example—your short-term memory will switch to that instead.

So other things, like a shopping list, might be lost.

There are ways to trick yourself into remembering! We call these tricks memory prompts.

Instead of struggling to retrieve information from your short-term memory, try using your episodic memory!

You don't remember the list itself, but do you remember the conversation you had with your father? In remembering that conversation, you might then remember the list!

This is an example of a **retrieval cue**. That's a memory association that helps you retrieve a specific memory. It's easier to remember the first thing on a list than stuff randomly in the middle. This is known as the **serial position effect**.

When you want to move something to your long-term memory, you have to pay attention to it! The best way to do that is through practice over a period of time. That tells your hippocampus that this is important.

When you sleep, your hippocampus replays what you want to remember over and over again, firming the synaptic pathways and making the memory stronger.

You'll probably forget a shopping list pretty quickly, but that time you were in a shop and a mob of clowns came in and scared you? You might remember that for the rest of your life.

OOP!

NAME: ~~ZOMBIE~~

1.) Who discovered the Pythagorean theorem and in what year?

Recall is the ability to retrieve what you learned or experienced previously.

PYTHAGORAS of GREECE, between 570 BCE & 495 BCE. It WAS likely discovered much eArlier IN ANCIENT BABYLONIA.

2.) Multiple-choice question: Which of the following is the Pythagorean theorem?
    a. $a^2 + b^2 = c^2$
    b. $a^3 + b^3 = c^3$
    c. $E = MC^2$

Recognition is the ability to identify information learned previously.

3.) Apply the Pythagorean theorem to determine the length of x:

$$6^2 + 8^2 = x^2$$

$$x = \sqrt{(6^2 + 8^2)}$$

Relearning is learning something for a second time. It goes more quickly than it did the first time.

Learning is the collection of knowledge or skills over time and the ability to adapt. You can examine past experiences and apply new information to solve problems.

Learning involves strengthening and changing synapses in your brain.

The stronger a synapse is, the faster you can react to a situation using the information, skills, or behavior that you've previously learned!

Learning is a lot like a path in a forest. The first time you walk the path, it can be hard to see where you're going.

You might not even know a path is there, but the more you walk it, the more clear and obvious the path becomes.

If this path is important enough to you, it might even turn into a street or highway! That's what learning is.

If you want to keep things really simple, there are two types of non-associative learning: **habituation** and **sensitization**.

Firstly, you might see something unusual and then get used to it over time.

SKREE!

Let's say that on your walk to school, you might come across a house with a huge green elephant in the yard.

Obviously, a giant green elephant would surprise anyone, and you might be very interested, but if that same elephant is there every day, you'll get used to seeing her.

Is that a green elephant?!

Yeah, her name is Eleanor Rosewood.

The more you're exposed to the same stimulus, the less you become interested in it. This is **habituation**.

INTEREST

TIME

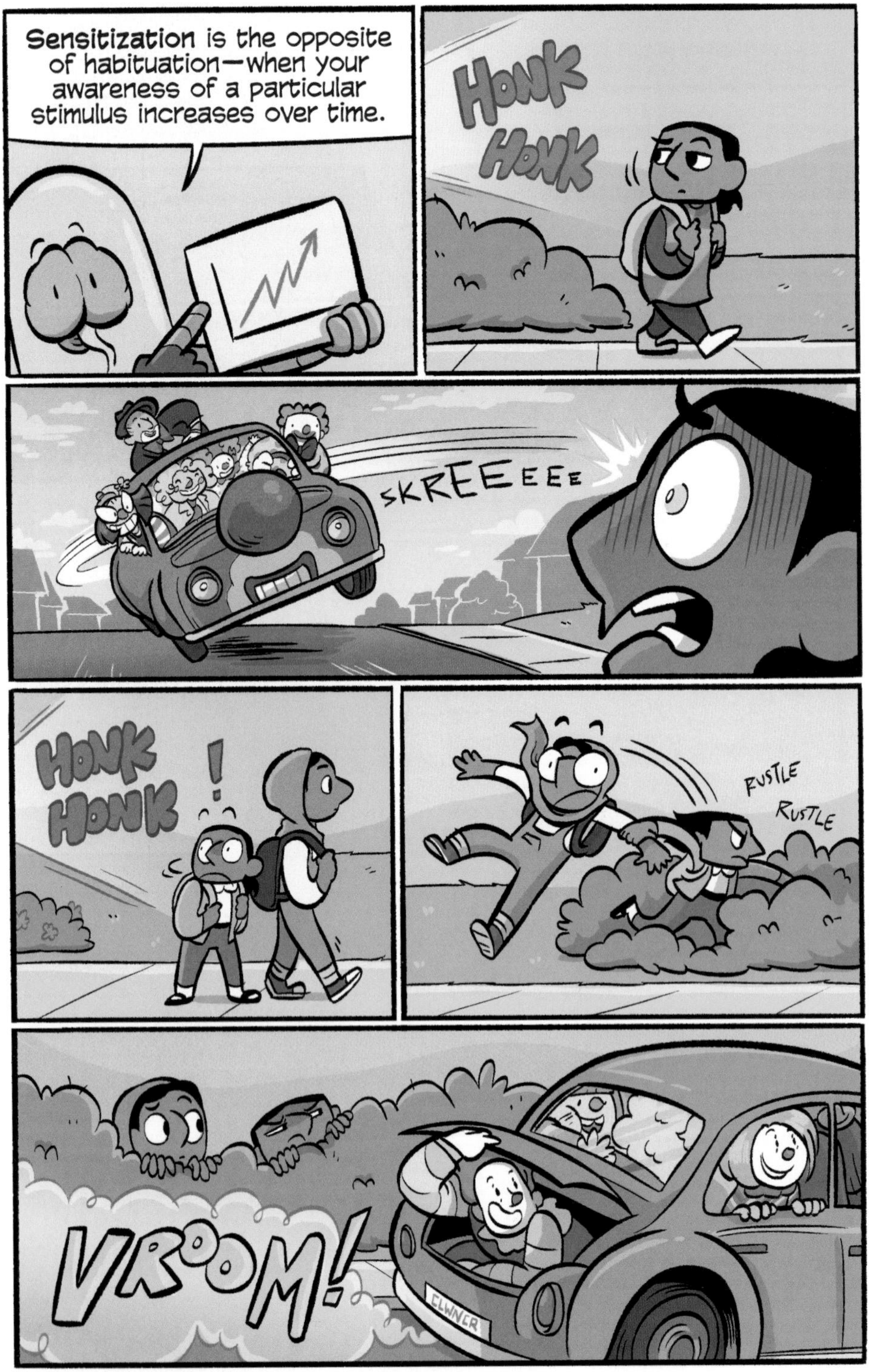
Sensitization is the opposite of habituation—when your awareness of a particular stimulus increases over time.
HONK
HONK
SKREEEEEE
HONK
HONK
!
RUSTLE
RUSTLE
VROOM!
CLWNCR

What about "proper" learning, like in a classroom or for a test?

Ahhh, you probably think studying for long periods at a time would be the best way to pass a test.

Isn't it?

Nope!

It's not about studying hard. It's about studying smart!

STUDY SMART!

Practice: Repeated practice over time is the best way to remember information! It's how you tell your brain that something is important and worth remembering.

The United States declared independence on July 4, 1776.

JULY 4, 1776

Pfft.

A simple way to do this is to use multiple methods: say it out loud, write it down, read it, and repeat it! *But...*

Study small: Study a little bit every day, instead of studying for hours all at once! This makes it easier for your brain to absorb and store the information. Think of it like moving 100 bricks. If you try to carry them all at once, it'll be too heavy, but carrying one brick at a time is much easier.

Sleep: Sleep, sleep, *SLEEP!* Sleep is super important in transferring knowledge from your short-term memory to your long-term memory. If you've studied small, it's *much* better to get a full night's sleep before a test than to stay up all night cramming!

So to recap (because repetition is important!): Repeat what you've learned at regular intervals, even if you think you already know it. Don't overload yourself. And get enough sleep!

That's not hard at all!

It's really not! And it'll make learning and remembering so much easier!

Think about intelligence like this: everyone has two eyes, a nose, and a mouth, but we all look very, very different.

Just like faces, no two brains are exactly the same! We all have two hemispheres, a corpus callosum, a temporal lobe, all that good stuff. But we all think, behave, and react differently.

The same is true for intelligence. No two people are exactly alike, and honestly, we would be in trouble if we all thought and behaved in exactly the same way! The world would end! Society would collapse!

That's a little dramatic.

Actually, not at all! To understand why, you have to understand what intelligence is.

Tell me, how would you define intelligence?

The ability to learn and acquire knowledge!

To be able to recognize problems and apply our knowledge to solve them.

The ability to communicate your ideas!

Teacher   Engineer   Artist

TYPES OF

Logic and Mathematics

A strong capacity to quantify, understand, and complete mathematical problems and to see patterns.

Linguistic

Use and understanding of language, both in speech and in thought. This is the ability to make yourself understood using language and the ability to understand language.

Spatial and Visual

The ability to visualize in three dimensions, to navigate the environment, and to reproduce mental images into a real-world space, i.e., drawing, sculpting, etc.

Kinesthetic

The ability to express one's feelings and ideas using your body. For example, the ability to score a goal in soccer or to express love and loss through dance.

There are basically eight types of intelligence.

Most people are a combination of these types, and not just exclusively one.

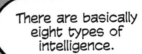

INTELLIGENCE

Naturalistic

The ability to understand the non-human environment, to recognize the differences between types of plants and animals, and to understand the natural world.

Musical

The ability to create, reproduce, and recognize the elements of pitch, rhythm, and tone.

Interpersonal and Social

The ability to understand the motivations of other people. To be able to interact with other people well, to be able to empathize with another person's perspective, and understand both verbal and nonverbal communication.

Intrapersonal

To be able to understand one's own goals and motivations. Being highly self-motivated and tending to think about existential questions, like "What's the meaning of life?"

(It's worth pointing out that this is just one theory of intelligence! There are all sorts of other theories out there. Everybody's brain is awesome!)

Every person in the world has a different mix of these types of intelligence!

For example, an architect needs to have very strong math and spatial intelligence.
They also need to be able to work alone, in addition to being able to work in a group.

Doctors need to have a strong understanding of biology and have the ability to communicate what's going on. They also need to be able to understand and empathize with their patients.

A farmer has to be able to work independently and be self-motivated. They need to understand the world around them in a literal sense, to weigh different factors like the wind, rain, and sun, and to apply these factors to different types of plants and animals with different needs.

Cartoonists need to be able to work alone and be self-motivated. They need to have an understanding of anatomy and the three-dimensional world in order to draw it. They also need to be able to use their body to help express themselves through drawing.

The mixture of different people with different types of intelligence is how we have innovation, invention, discovery, and society!

Some people have a kind of intelligence that lets them do really well in school, but there are also plenty of people who have a kind of intelligence that lets them do really well in sports, drawing, or any other number of skills! Don't ever underestimate your intelligence or the intelligence of those around you!

Well, that was a great speech. Very inspiring! I'm going to go use all this new-found knowledge to make myself a better person!

Seeya!

OH-HO-HO!

Not so fast, my dear!

There's still the matter of that precious, clever little brain of yours...

But...but...why do you even want my brain anyway?!

We get lonely down here.

Now hold still, you won't feel a—

DINGDONG!

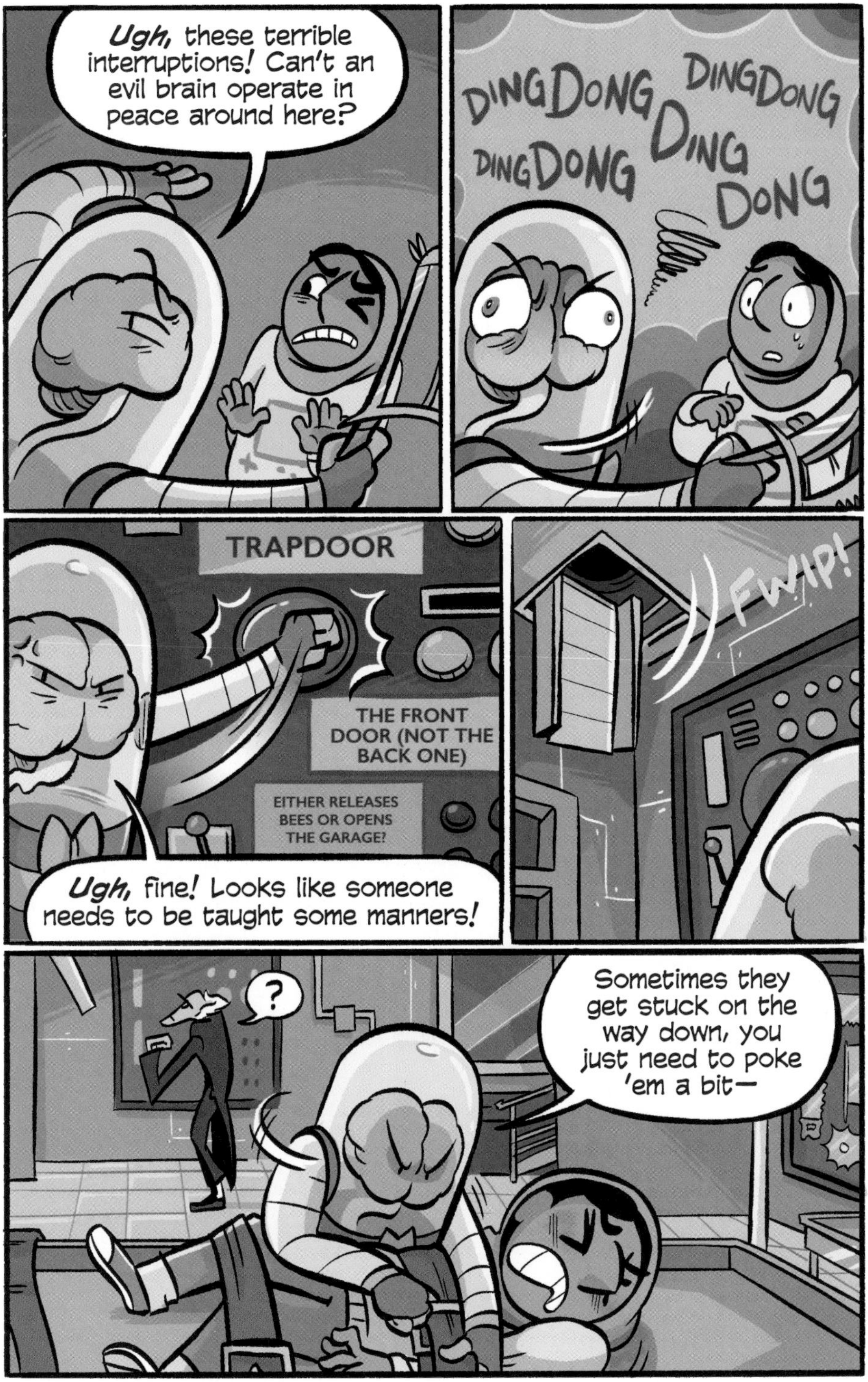
Ugh, these terrible interruptions! Can't an evil brain operate in peace around here?
DING DONG DING DONG DING DONG DING DONG
TRAPDOOR
THE FRONT DOOR (NOT THE BACK ONE)
EITHER RELEASES BEES OR OPENS THE GARAGE?
Ugh, fine! Looks like someone needs to be taught some manners!
FWIP!
?
Sometimes they get stuck on the way down, you just need to poke 'em a bit—

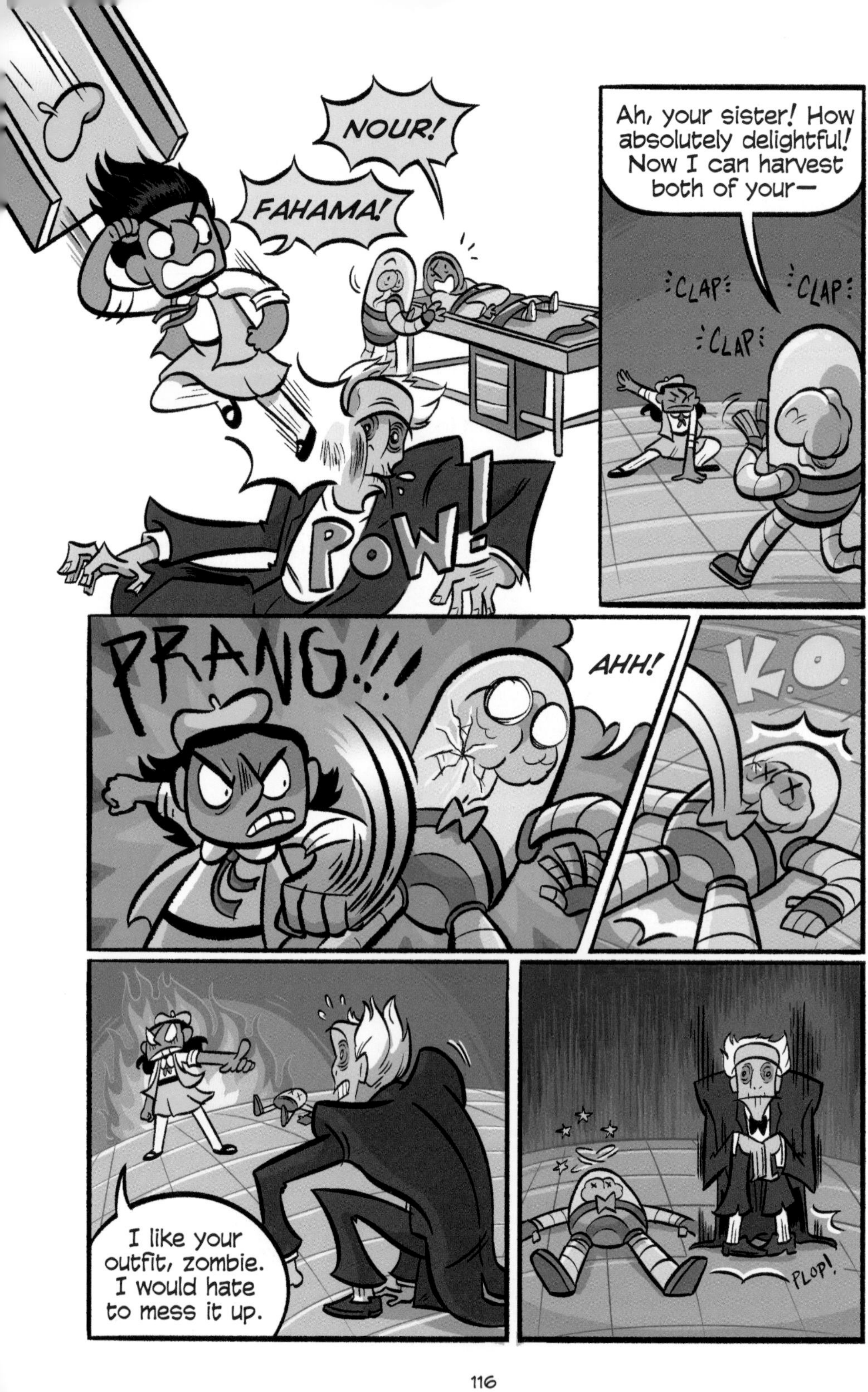
FAHAMA!
NOUR!
POW!
Ah, your sister! How absolutely delightful! Now I can harvest both of your—
CLAP
CLAP
CLAP
PRANG!!!
AHH!
K.O.
I like your outfit, zombie. I would hate to mess it up.
PLOP!

Some time later...

DING DONG

Would you like to buy some cookies to support our extracurricular activities?

These cookies are made with...

Braaains?

No, of course not!

Best sister in the world

117

—GLOSSARY—

Action potential

A change in polarity across the membrane of a neuron's axon, action potential is what releases neurotransmitters, allowing neurons to talk to one another.

Axon

A projection off a nerve cell that the action potential travels down.

Bilateral symmetry

Organisms that are divided into two symmetrical halves, left and right. Humans, birds, and worms are all examples of bilateral symmetry.

Brain stem

Helps to run most automatic bodily functions (the things your body does without having to think about it), like breathing and keeping your heart beating!

Capillary

The smallest type of blood vessel. Capillaries form networks in your organs to transport oxygen and nutrients to the body's cells and to help remove waste.

Cell

The smallest unit of a living thing.

Cerebellum

Located at the back of the human brain, the cerebellum tells the muscles of the body how to balance and move.

Cerebrum

Comprising two-thirds of the human brain, the cerebrum is the largest part of the brain. The cerebrum is where decision making, imagination, and language happen!

Dendrite
Branch-like projections coming off the cell body of a neuron that receive signals from other cells.

Diencephalon
The part of the human brain that contains the thalamus, hypothalamus, pituitary gland, and pineal gland. It helps to regulate your hormones and other bodily functions.

Excitatory
To increase activity or excitement, such as in a neuron.

Excitatory neurotransmitters
Chemicals your neurons release that increase the activity of the receiving neuron.

Habituation
A type of non-associative learning where the more you're exposed to a stimulus, the less you respond to it.

Homeostasis
When your body is in a state of order and everything is normal. It's how your body likes to be most of the time—calm and relaxed.

Inhibitory
To decrease activity or prohibit chemicals from being received by a neuron.

Inhibitory neurotransmitters
Chemicals your neurons release that decrease the activity of the receiving neuron.

Instinct
A behavior in animals and people that isn't learned. For example, babies are born instinctively knowing how to suckle.

Insulation

Material that is added to an object to prevent the loss of energy. For example, insulating your home prevents the loss of heat, and insulating your axons prevents the loss of action potential!

Lateral

The side of something. In the case of an animal or a human, it often refers to the sides of an animal ("left lateral" or "right lateral").

Lobe

A clear anatomical division of an organ. In the case of the human brain, the four lobes of the cerebral cortex are the frontal lobe, the parietal lobe, the occipital lobe, and the temporal lobe.

Neurons

A cell that is specialized to send and receive nerve impulses.

Neurotransmitter

Chemicals released by neurons, and are used by brain cells to communicate with one another and the rest of the body.

Neocortex

The newest part of the brain to evolve, the neocortex is the outermost part of the brain. The neocortex is made up of six layers of brain cells on the outermost part of the cerebrum.

Nerves

A collection of axons that transmit information to and from different parts of the body.

Node

A central point where things connect to each other.

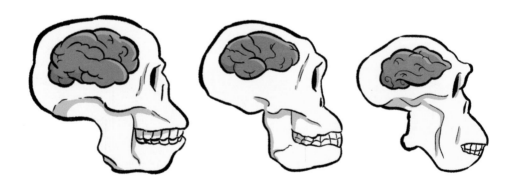

Non-associative learning
> Learning from repeated exposure to stimulus.

Olfaction
> The sense of smell. For example, "olfactory sensors" refers to receptors related to smell. Adapted from the Latin word *olfactus* meaning "smelled."

Organelle
> A tiny specialized structure in a cell, similar to an organ in your body. It helps keep the cell functioning normally.

Perception
> How you organize, interpret, and contextualize the information received from stimuli.

Radial symmetry
> Appears in organisms that don't have a distinct left and right side, but rather have symmetry around a central point. Sea anemones, jellyfish, and starfish are all examples of radial symmetry.

Reflex
> An action where the body reacts to stimuli without conscious thought.

Sensitization
> Sensitization is a type of non-associative learning where your awareness of a particular stimulus increases over time.

Synapse
> The point where two neurons meet and neurotransmitters pass between them.